IMAGES
of America

Flaming Gorge Dam

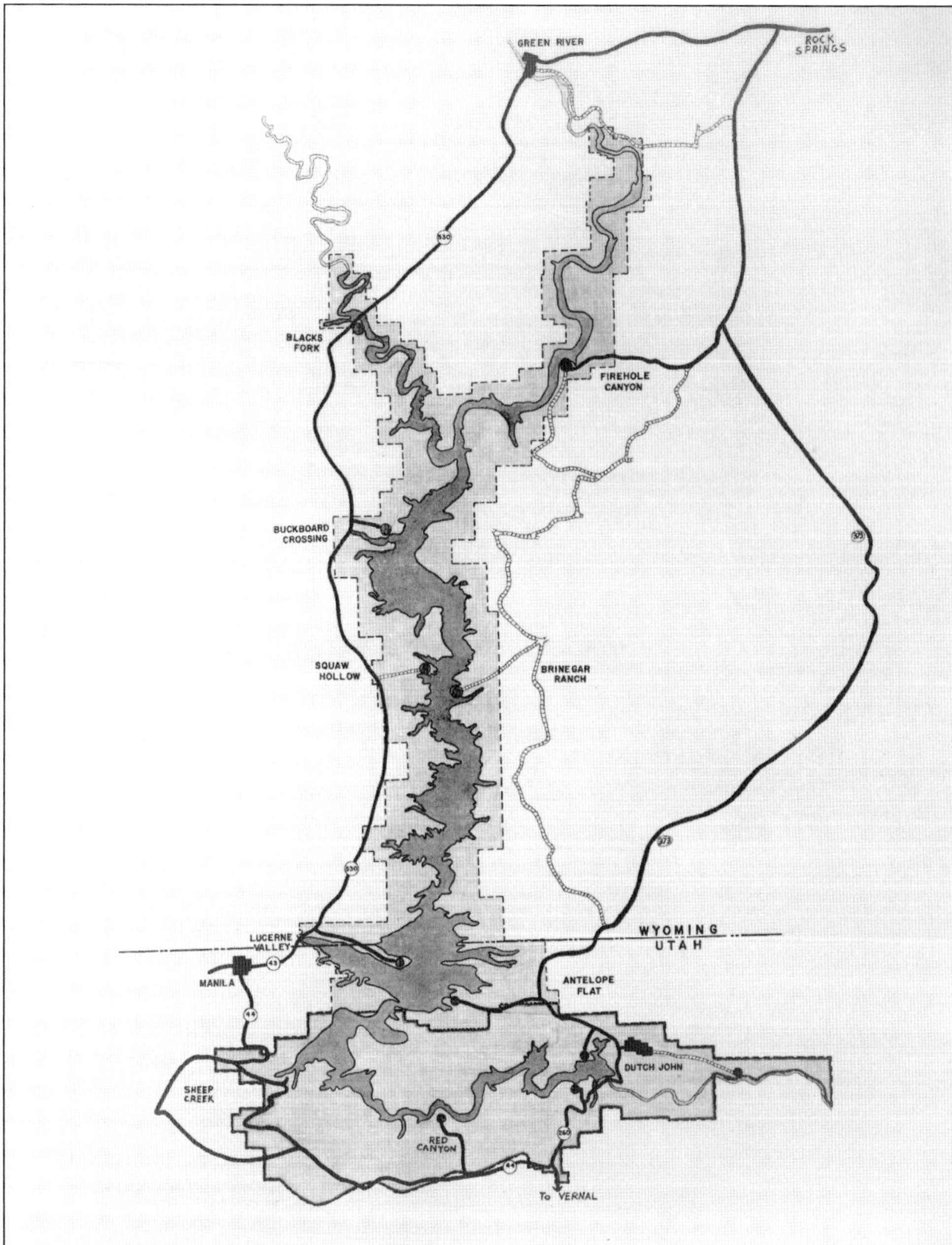

The Flaming Gorge Reservoir was formed from the Green River as a result of construction of the Flaming Gorge Dam. The reservoir is 91 miles long and has a depth of 436 feet at the dam. Recreation areas dot the 375-mile perimeter for swimming, boating, camping, and fishing. (Courtesy of US Forest Service.)

On the Cover: Joe V. Sanchez stops to watch the concrete bucket (upper left) lower to empty concrete onto the dam. Concrete has been placed in the blocks to an average of 300 feet above the foundation. Approximately 29 percent of the dam is concealed by water on the reservoir side of the structure in this photograph taken on May 23, 1962. (Courtesy of Bureau of Reclamation.)

IMAGES
of America

FLAMING GORGE DAM

Uintah County Regional History Center

ARCADIA
PUBLISHING

ISBN 978-1-4671-3016-5

Published by Arcadia Publishing
Charleston, South Carolina

Printed in the United States of America

Library of Congress Control Number: 2013945231

For all general information, please contact Arcadia Publishing:
Telephone 843-853-2070
Fax 843-853-0044
E-mail sales@arcadiapublishing.com
For customer service and orders:
Toll-Free 1-888-313-2665

Visit us on the Internet at www.arcadiapublishing.com

Dedicated to the Upper Colorado River Storage Project Board and its members, who fought so hard and for so long to have a dam built on the Green River.

Contents

ACKNOWLEDGMENTS

The Uintah County Regional History Center staff would like to thank Doris Karren Burton for her vision and work to begin this wonderful center, which is said to be the best in the Western region and the only one of its kind in the state of Utah. Doris aggressively and tirelessly found funding and donations to increase the center's historical archives. Thank you to all other volunteers and staff who have contributed to the growth and preservation of our regional history.

Thanks go to Sam Passey, library director, who encourages us in our work, and to the Uintah County Commissioners, who support us and allow us to have this wonderful job of sharing our love for history with our patrons. We would like to recognize and give thanks to Lee Cheeves and the Uintah County Library Board for their encouragement in our behalf.

We appreciate the friendly and giving personnel of the Bureau of Reclamation—Ginger Reeve, Pauline Baker, and Roxanne Reid—who generously donated photographs for this book. Thank you to Christopher Nichols and the US Forest Service in Dutch John, who allowed us to scan and use anything they had available. Thanks so much to Vicky McKee, Daggett County clerk treasurer, and Melina Coleman of the Daggett County Museum for all their help in providing photographs. Thank you to Cynthia McCullers of the Sweetwater County Museum in Green River, Wyoming, for the use of several photographs.

Many photographs are housed in our center, including the Vernal Express and Leo Thorne collections, which were used for this publication. The staff of the Uintah County Regional History Center includes Ellen Kiever, Elaine Carr, Michelle Fuller, and Linda Wilson.

The majority of images in this publication appear courtesy of Bureau of Reclamation (BOR), US Forest Service (USFS), and Uintah County Regional History Center (UCRHC). The sources for all other images are credited in full.

INTRODUCTION

Trapper and geologist John Wesley Powell traveled the Green River in 1869, looking for beaver and other fur animals. He observed from his raft the brilliant red gorge ahead. He said the mountain ridge through which the gorge cut was composed of bright vermilion rocks and surmounted by broad bands of mottled buff and gray that curved down to the water's edge. He stated, "We name it Flaming Gorge." Powell noted that the future of the Mountain West depended on water resources.

The Green River, the main tributary to the Colorado River, had been a major topic of discussion by organized water project groups since the early 1920s. The most favorable solution was to build a dam at Flaming Gorge. Surveys were conducted in the canyon, proposals were made, and talks on the subject continued, but that was all that became of it. This kind of discussion carried on and was printed in newspapers for the next 25 years.

In the late 1940s, the Upper Colorado River states were all concerned about getting their share of the water. In 1949, the Bureau of Reclamation submitted a report on the Colorado River Storage Project; from this, the Upper Colorado River Storage Project was formed. Committee members from each of the Upper Colorado River States worked diligently to draw up the best plan for water storage on the Green River. Three possible sites were submitted for a dam site.

Finally, after 35 years of talk about Flaming Gorge as a dam site, it was made official. On October 15, 1956, Pres. Dwight Eisenhower gave the signal from Washington, DC, and a charge of dynamite lifted rocks high into the air. Those in attendance watched with a feeling of respect and gratitude as smoke filled the clear blue sky. There were lumps in many throats as officials realized that, through unselfish efforts, the ground-breaking of one of the nation's largest river projects was at last a reality. Before completion of the dam in 1964, three American presidents would play a vital part in the Flaming Gorge Dam project.

When construction workers began arriving to work on the dam, it was necessary to create a new town near the construction site for them to live. The town, built on Dutch John Flat, was named Dutch John. The Arch Dam Construction Company brought in trailers for its workers. Later, a bunkhouse was built and a hospital made available. With the workers came their families, and more houses were built, along with a school, store, and post office.

A new town was created, but the demise of an old town was in the near future. Linwood, Utah, situated five miles east of Manila on Henry's Fork, was located at an elevation below which the Green River would rise into a large reservoir. The town's homes and stores were either moved or burned, and the land was cleared of debris. Later, the old town was buried by water.

In October 1958, the Utah State Road Commission approved the contracts for the roads and bridges that would link Dutch John with Vernal, Utah. This new construction would save approximately 45 miles in driving distance for the community and workers in the area. A suspension bridge was built across the Green River between Angel Falls and the dam site. It provided a one-way pass for use of the workers and commuters. The Cart Creek Bridge was constructed with two massive arches to support the decking on which the cars would pass. It spanned a 300-foot-deep canyon. When the Cart Creek Bridge was complete, the suspension bridge was removed, and the road construction from the Manila Highway to the dam was finished. It is a beautiful, scenic drive down to the dam and across to the town of Dutch John. A cement road, built against the mountain on the northeast side of the dam, was unique for roads in the area.

Before the dam could be constructed, the excavation phase was undertaken. This included drilling a 1,100-foot tunnel through a mountain to divert the Green River from the construction site. A cable system was strung across the canyon, 500 feet above the river, to deliver buckets of concrete to the dam structure. An aggregate plant was built near Linwood, and a concrete batch plant was constructed near the dam, complete with a short-line railroad to deliver the concrete to the large bucket.

The first bucket of concrete was poured into a block of the dam on September 8, 1960. Bucket by bucket and block by block, the arch dam structure began to rise. The final bucket of concrete was placed on November 15, 1962, making the dam structure 502 feet high. Although this portion of the project was complete, there was still much to do.

A powerhouse was built on the downstream side of the dam to house three turbine generators. Each generator provides 36,000 kilowatts of electricity for a total of 108,000 kilowatts. The powerhouse then generates the power to the switchyard west of the dam, where electricity flows through the lines to distant cities, providing power to thousands of homes.

Pres. John F. Kennedy came to Utah and participated in a brief ceremony in which he pressed the key that started the first generator at the Flaming Gorge powerhouse. First Lady, Lady Bird Johnson, flew into Vernal, Utah, where she was greeted by a large crowd of local citizens who welcomed her with pride and respect. She traveled by bus up the Vernal-Manila Highway to the completed dam, where she was joined by other dignitaries and an excited group of spectators. She was the first woman to dedicate a project for the Bureau of Reclamation.

The coming of the dam made quite a difference to Manila, Utah. Many people in the valley worked on the construction of the dam. For some, it made the difference between bare subsistence and a good living. The former isolated ranching community soon became a tourist destination, with new restaurants, motels, campgrounds, trailer courts, and even a supermarket. Service stations now cater to boats as much as they do to automobiles. The town's young people no longer had to go elsewhere to find a job; they could find employment in Manila with a government agency or in the tourism industry.

The Flaming Gorge National Recreation Area, surrounded by brightly colored canyon walls, is set amid hundreds of thousands of acres of forested hills. It is a wonderful setting in which to boat, fish, hike, bike, or just stare. The Green River below the dam provides some of the best trout fishing in Utah and Wyoming, as well as mild white-water rafting. Dozens of campgrounds have been developed by the forest service, along with five full-service marinas to launch boats and other watercraft. Each year, tourists from all over the world visit the Flaming Gorge Dam and Recreation Area.

One

It Began with a Bang

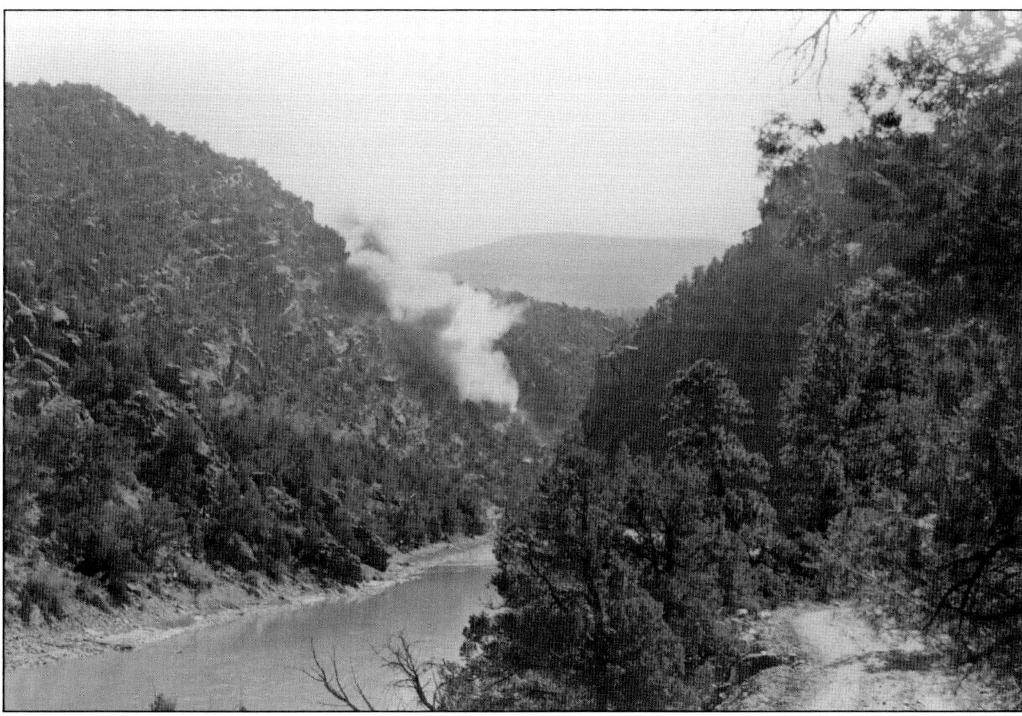

On October 15, 1956, at a signal given by Pres. Dwight Eisenhower in Washington, DC, a blast of dynamite filled the clear blue sky with dust and rock particles, marking the inauguration of the Flaming Gorge Dam. Over 80 persons in attendance watched with emotion as the event unfolded after years of hard work and effort to get this large river project off the ground. (Courtesy of UCRHC.)

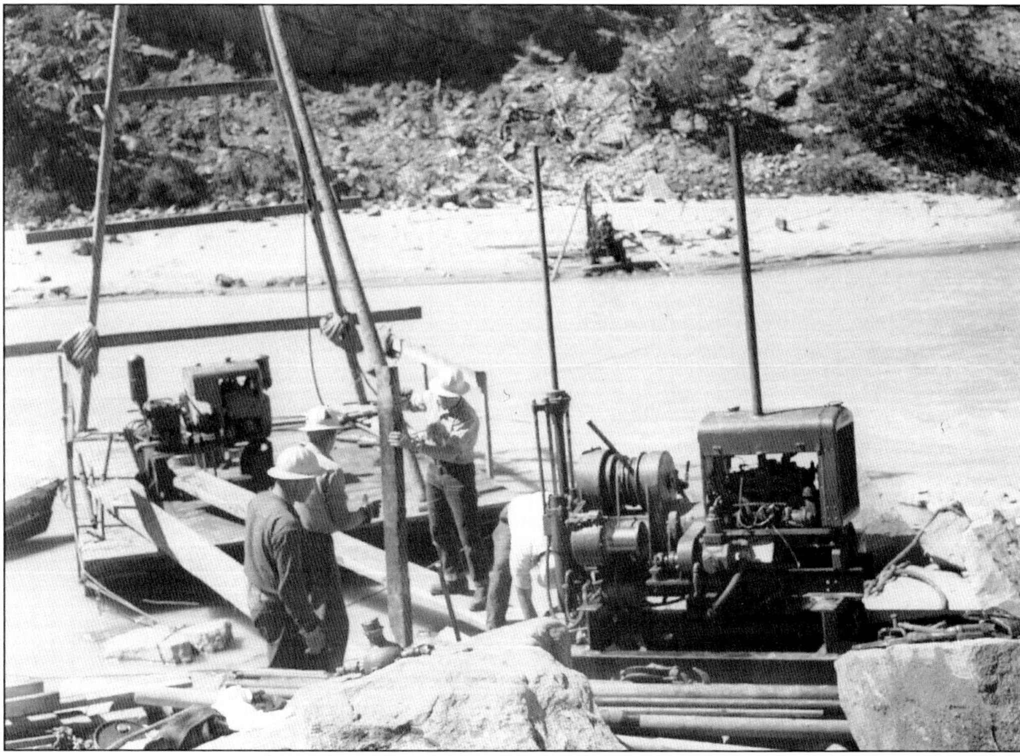

In September 1956, drilling barges built at the Flaming Gorge Dam site were used to determine rock conditions for the foundation of the dam. Drilling rigs were installed on the barges, which were anchored at various drilling sites in the swift-flowing river. Casing was set in the river until rock was encountered. In two shifts, 25 men worked under the supervision of George Martinsen, foreman of the barge crew. An explosion on one of the barges threw two men into the river and severely burned Clinton Gerber Price on his face. (Courtesy of UCRHC.)

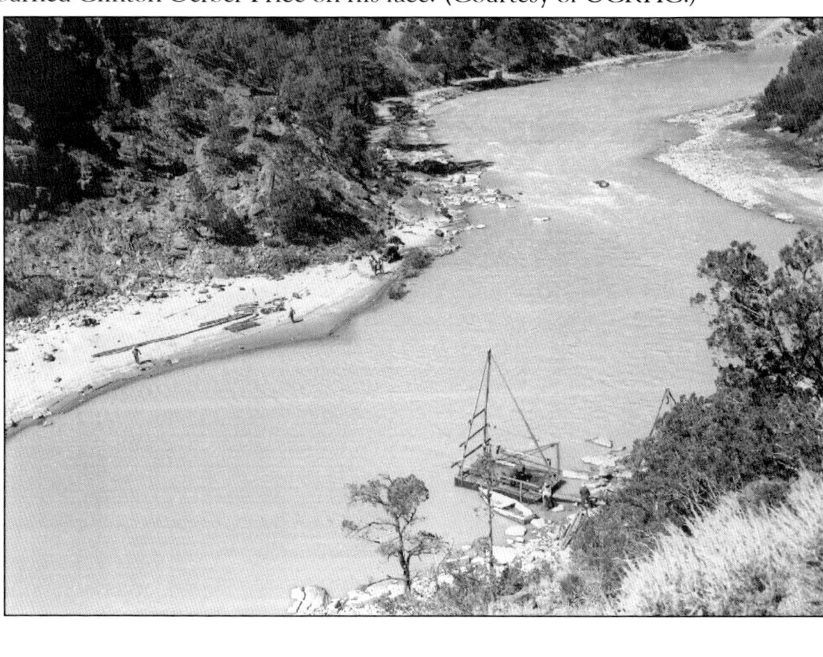

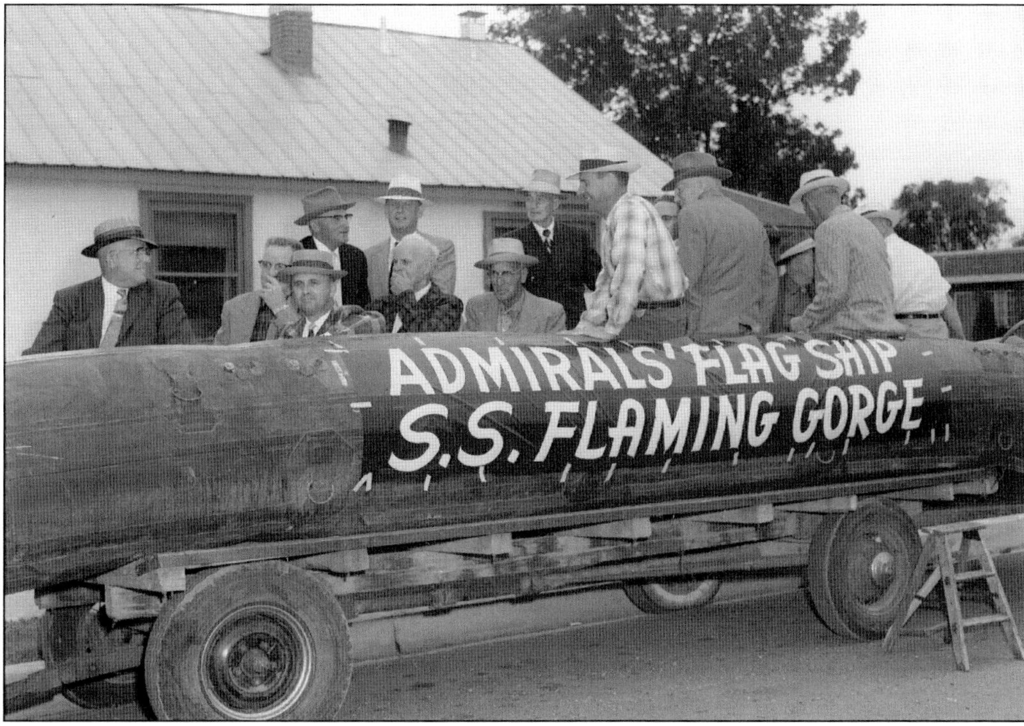

The biggest parade in the history of Vernal drew large crowds as the enthusiastic community celebrated congressional appropriation of $1 million for the Vernal Flaming Gorge Project. Above, leading the parade in one of Bus Hatch's riverboats is Gov. George D. Clyde (far left), with members of Utah Water and Power Board. Sitting behind Governor Clyde are, from left to right, Gordon R. Clark (representing Saint George), J.R. Bingham (Salt Lake City), Wallace Yardley (Beaver), Charles Reed (LaSal), Dr. P.L. Jones (Nephi), Raymond Hammond (Tooele), J.P. Stevens (Henefer), Heber Weingate (Monroe), Wayne D. Criddle (Salt Lake City), Byron O. Colton (Roosevelt), David Scott (Ogden), Leo Harvey (Pleasant Grove), and Orville Lee (Paradise). Shown in the photograph below are, from left to right, Horace Allred, Hugh Colton, Governor Clyde, Byron O. Colton, and Wayne Criddle. (Courtesy of UCRHC.)

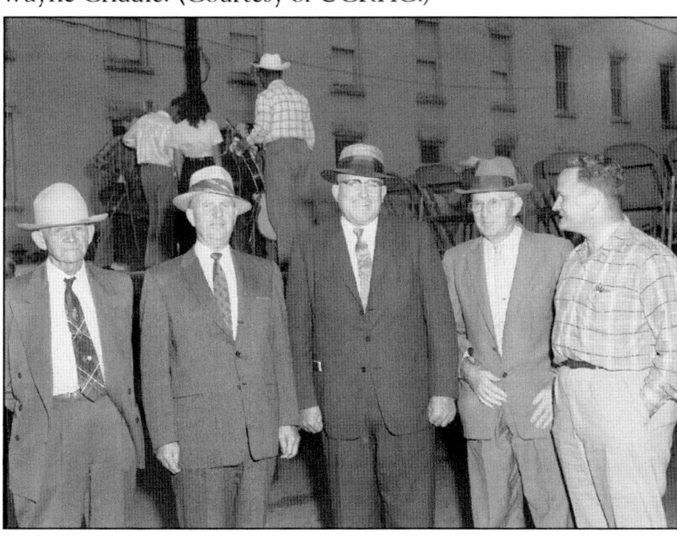

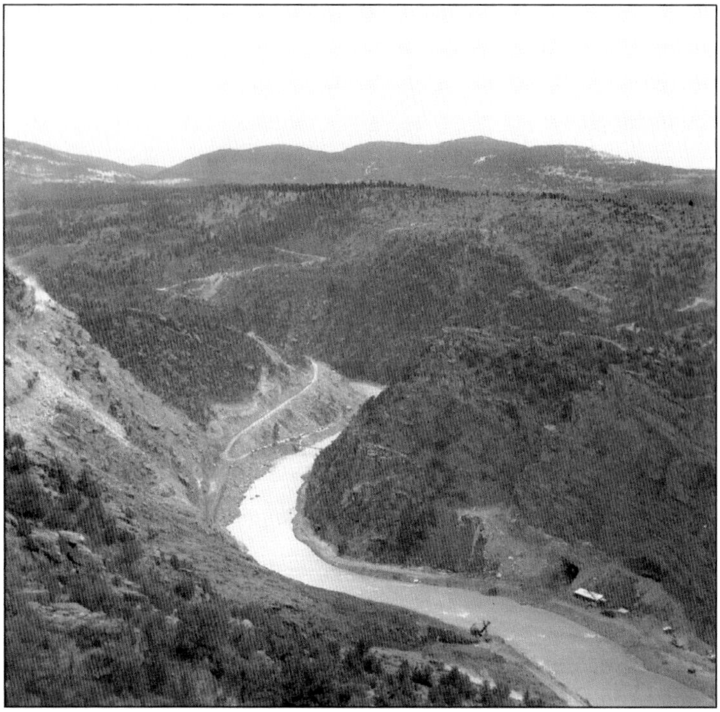

A breathtaking view of Flaming Gorge is seen from the lookout point, under construction near Dutch John. The diversion tunnel is in the lower right of the photograph. To the left center is the road, under construction, that winds down to the Green River. (Courtesy of UCRHC.)

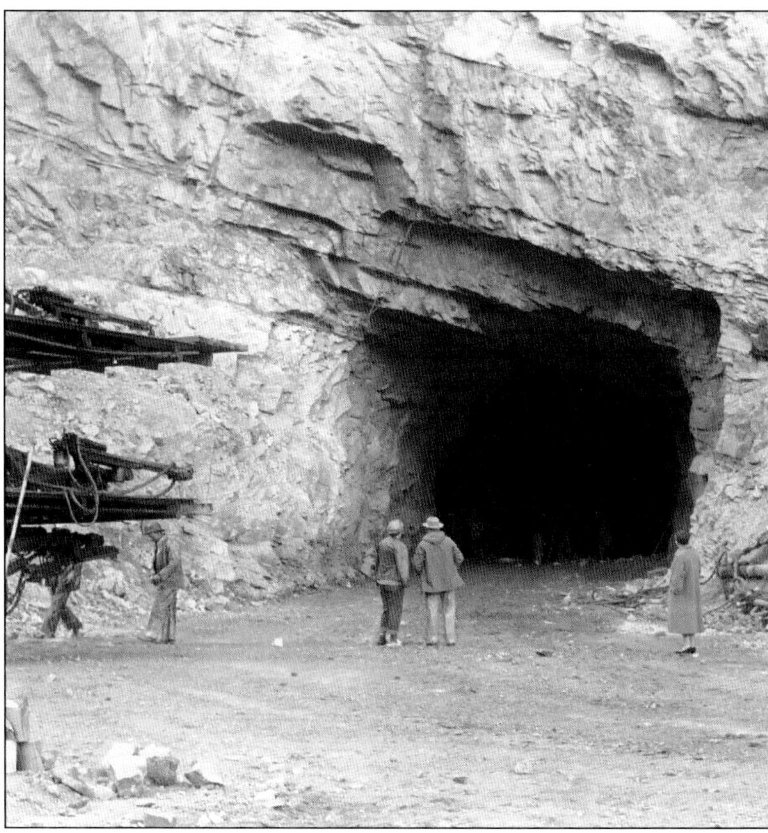

The diversion tunnel, seen here on January 15, 1959, appears as a gaping black hole in the rock ledge at the bottom of Flaming Gorge Canyon. The tunnel, located on the east side of the Green River above the dam site, was the first major phase of the Flaming Gorge Dam construction. At the time of this photograph, the tunnel had been driven 80 feet into the rock surface. When finished, the tunnel diverted the Green River away from the dam site. (Courtesy of UCRHC.)

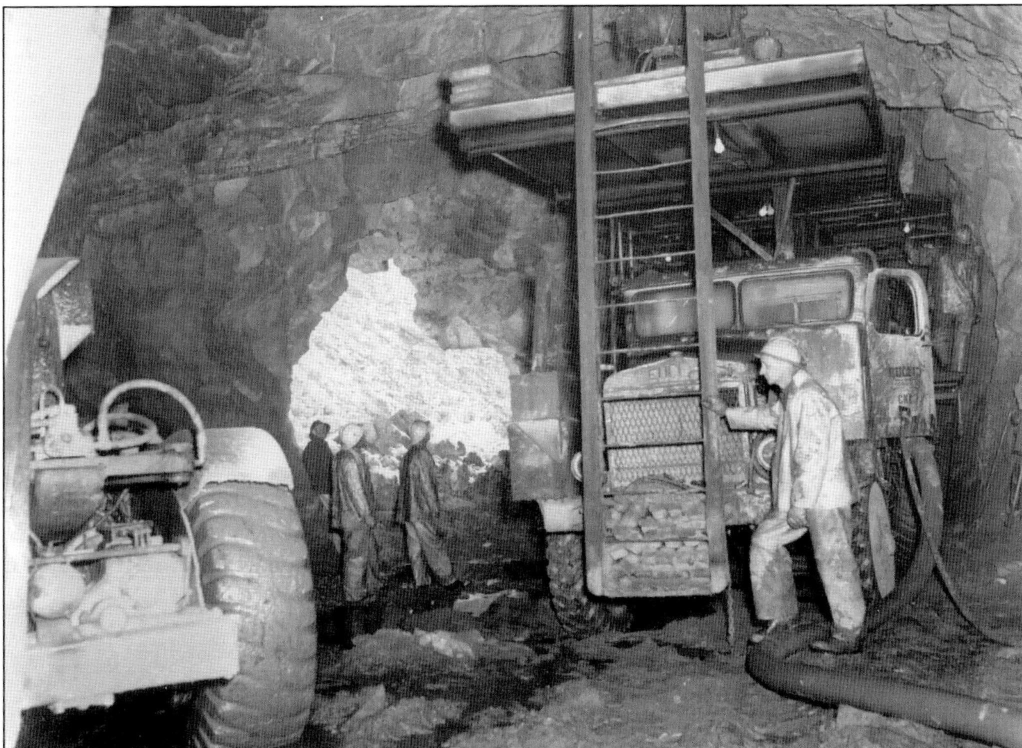

The sun looks good to crew members (below) after they break through the mountain of rock in the 1,100-foot-long diversion tunnel on March 12, 1959. Unidentified men are located inside the tunnel with heavy construction equipment (above). The 23-foot-wide tunnel will be lined with a three-foot-thick concrete layer. The Green River is scheduled to be diverted by September 1, shortly after the runoff season. (Courtesy of BOR.)

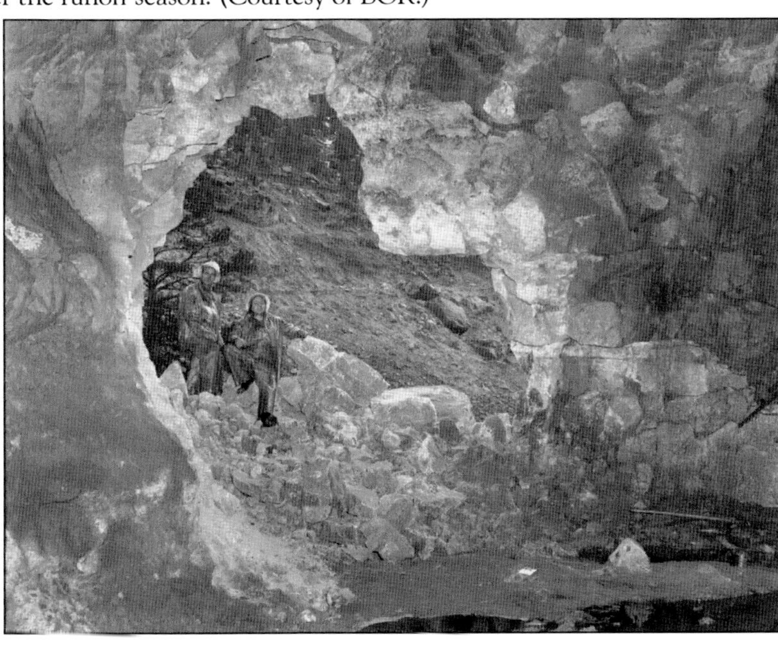

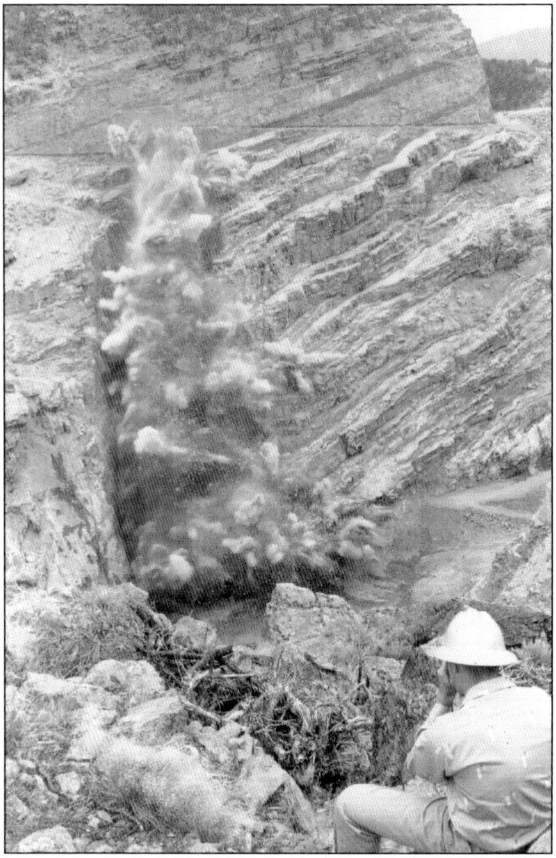

Civil engineer Frank Peterlin watches as a blast is made in the left keyway of the Flaming Gorge Dam site on July 7, 1960. The trimming cut was made to shape the keyway to the proper contours for the foundation of the concrete arch structure, which will be approximately 500 feet above the rock foundation. (Courtesy of BOR.)

Simultaneous blasts were set off in the right and left keyways of the excavation site of the Flaming Gorge Dam on July 21, 1960. The shot on the right abutment was the final blast on that portion of the excavation, and the blast on the left abutment was for the river outlets. The cofferdam, seen in the center, holds back the waters of the Green River as it flows through the diversion tunnel. (Courtesy of BOR.)

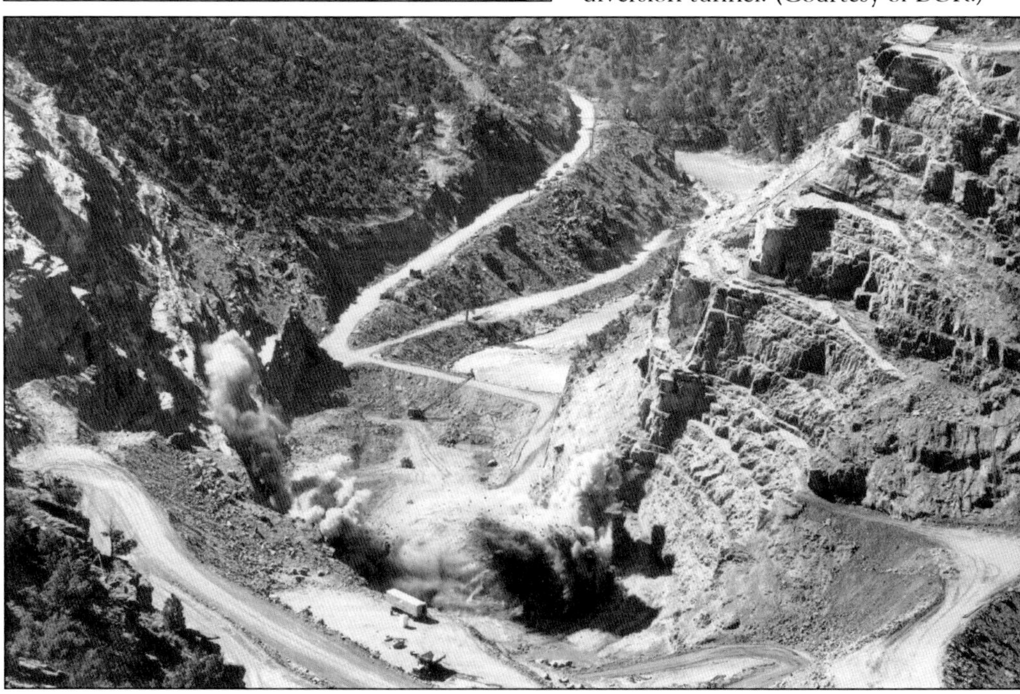

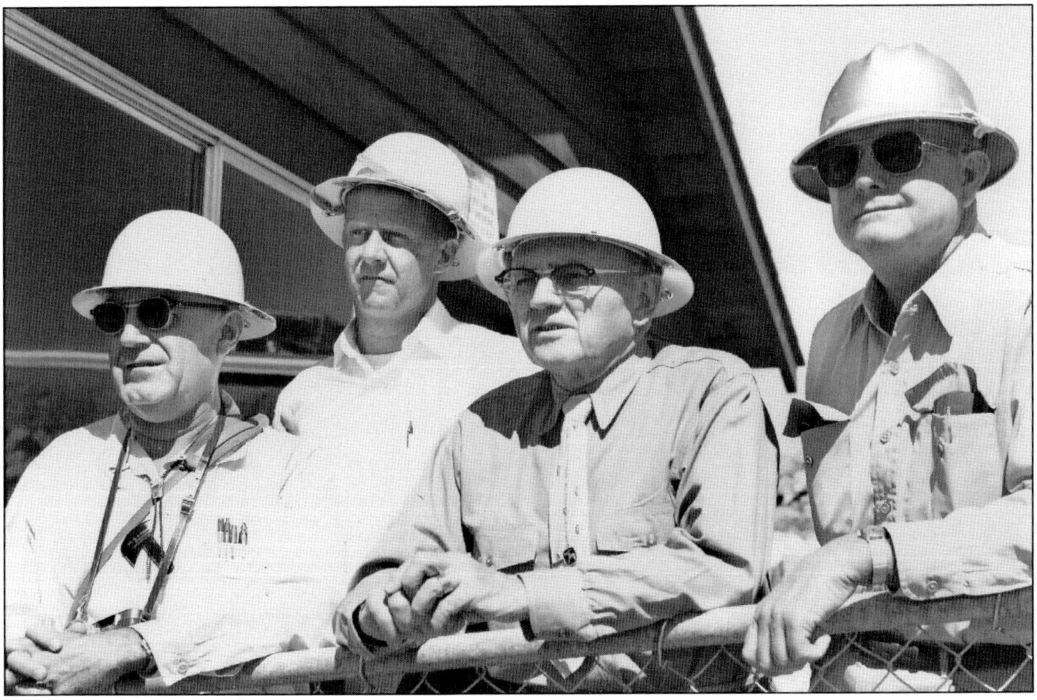

In September 1960, dignitaries visited the Flaming Gorge Dam project to inspect the progress of the work. They are, from left to right, Colorado Water Conservation Board member William H. Nelson, Tom Neal, Congressman Wayne N. Aspinall, and Bureau of Reclamation project engineer Jean R. Walton. (Courtesy of UCRHC.)

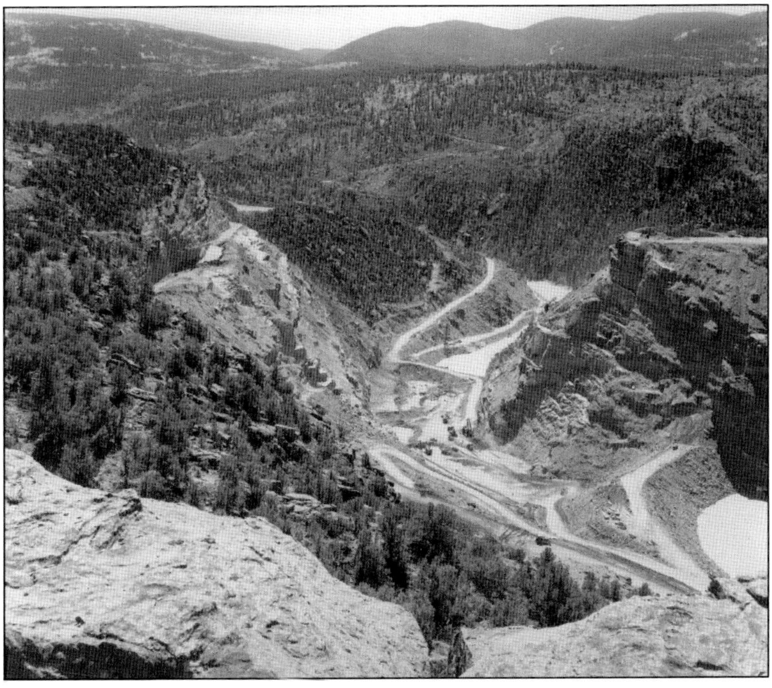

In November 1959, a cofferdam was completed on each side of the Flaming Gorge Dam site, holding back the Green River during construction. Each dam was approximately 260 feet long, 80 feet through the base, and 23 feet above the water level. Vehicles and equipment can be seen where the dam will be located. (Courtesy of BOR.)

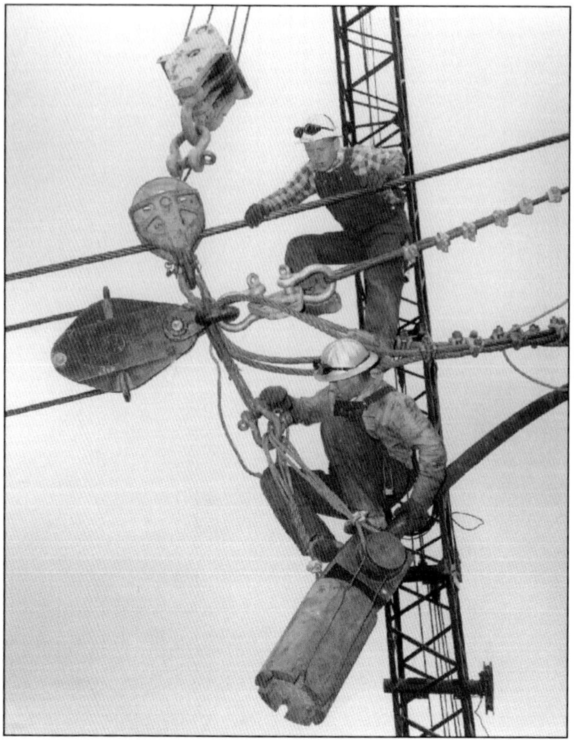

Ray E. Adams (top) and his father, Frank Adams, hang like monkeys on the cable line crossing Flaming Gorge Canyon. They are shown shackling the main cable to the 1 1/2-inch cable in July 1960. (Courtesy of BOR.)

In July 1960, the first main cable is shown on its way across the canyon at the Flaming Gorge Dam development. The two 3 1/4-inch cables will support eight-cubic-yard buckets for placing concrete on the dam. Each cable supports a working load of 25 tons. The men are unidentified. (Courtesy of BOR.)

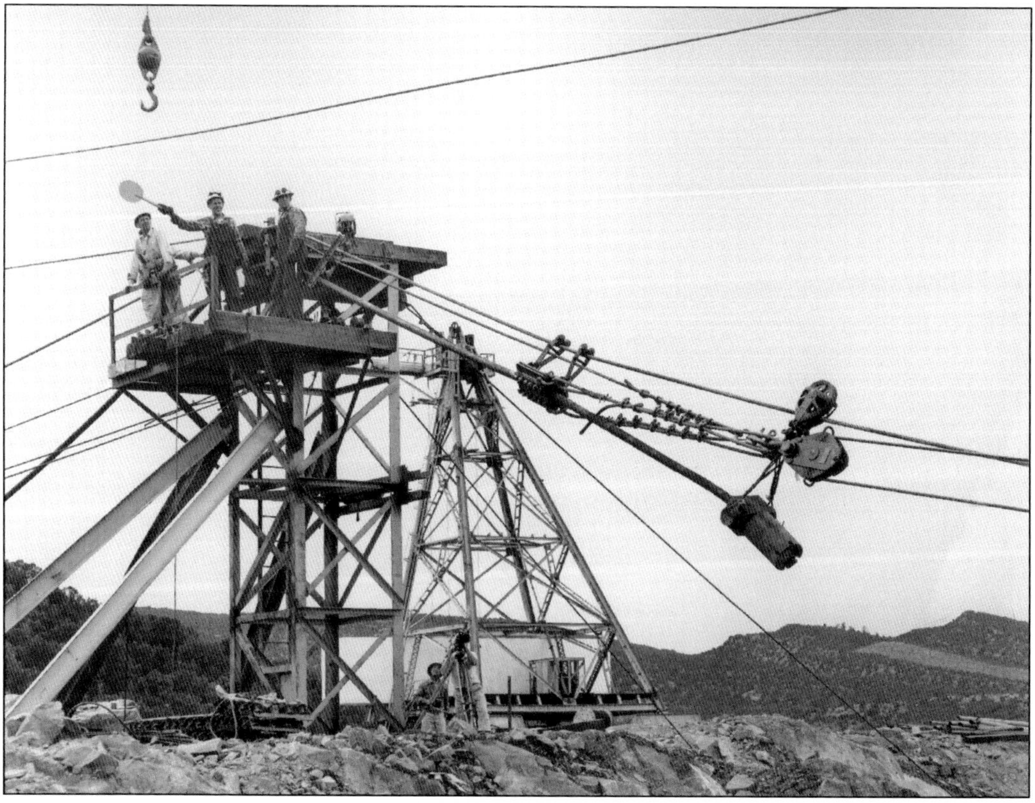

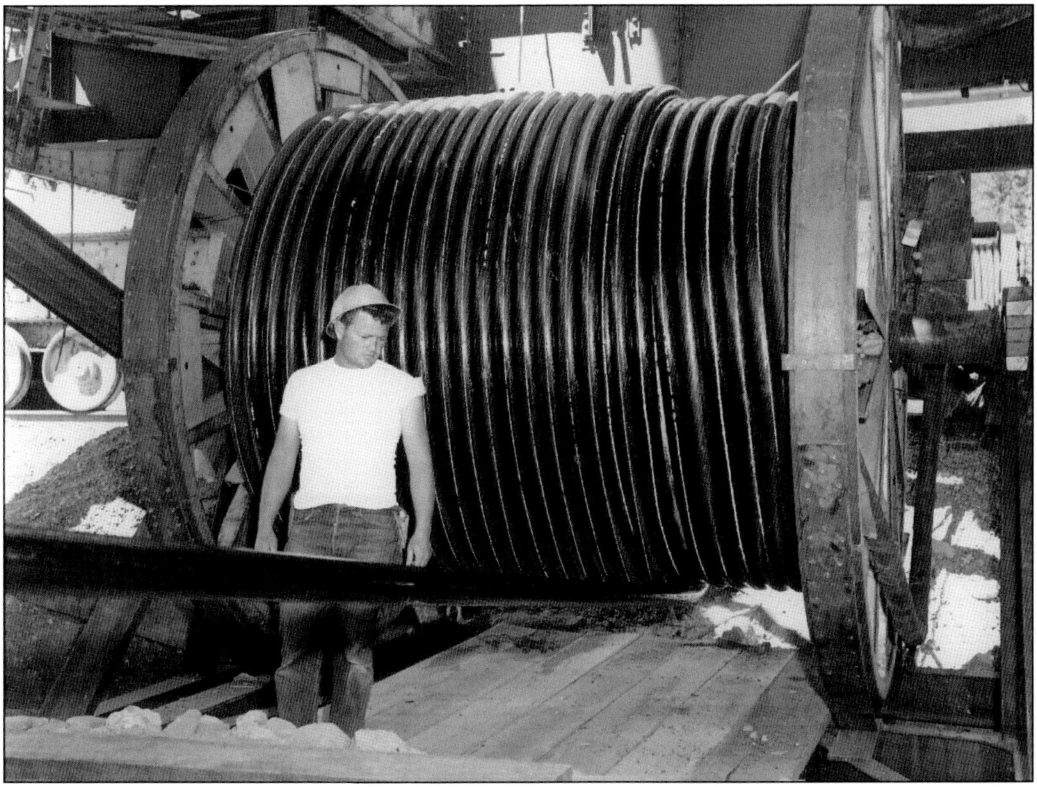

D.A. Mitchell of Manila inspects the main cable reel as it is unrolled during the stringing operation of the cableway system across the canyon at Flaming Gorge Dam in July 1960. (Courtesy of BOR.)

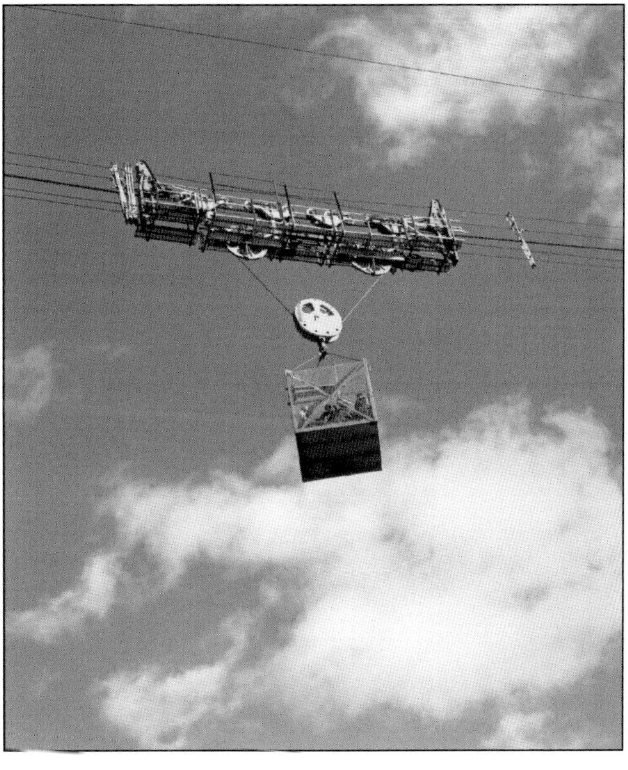

Workers on the dam sometimes found it necessary to take the high road to work. Men are seen riding the personnel cage on the cableway to the right abutment of the dam on March 13, 1961. (Courtesy of BOR.)

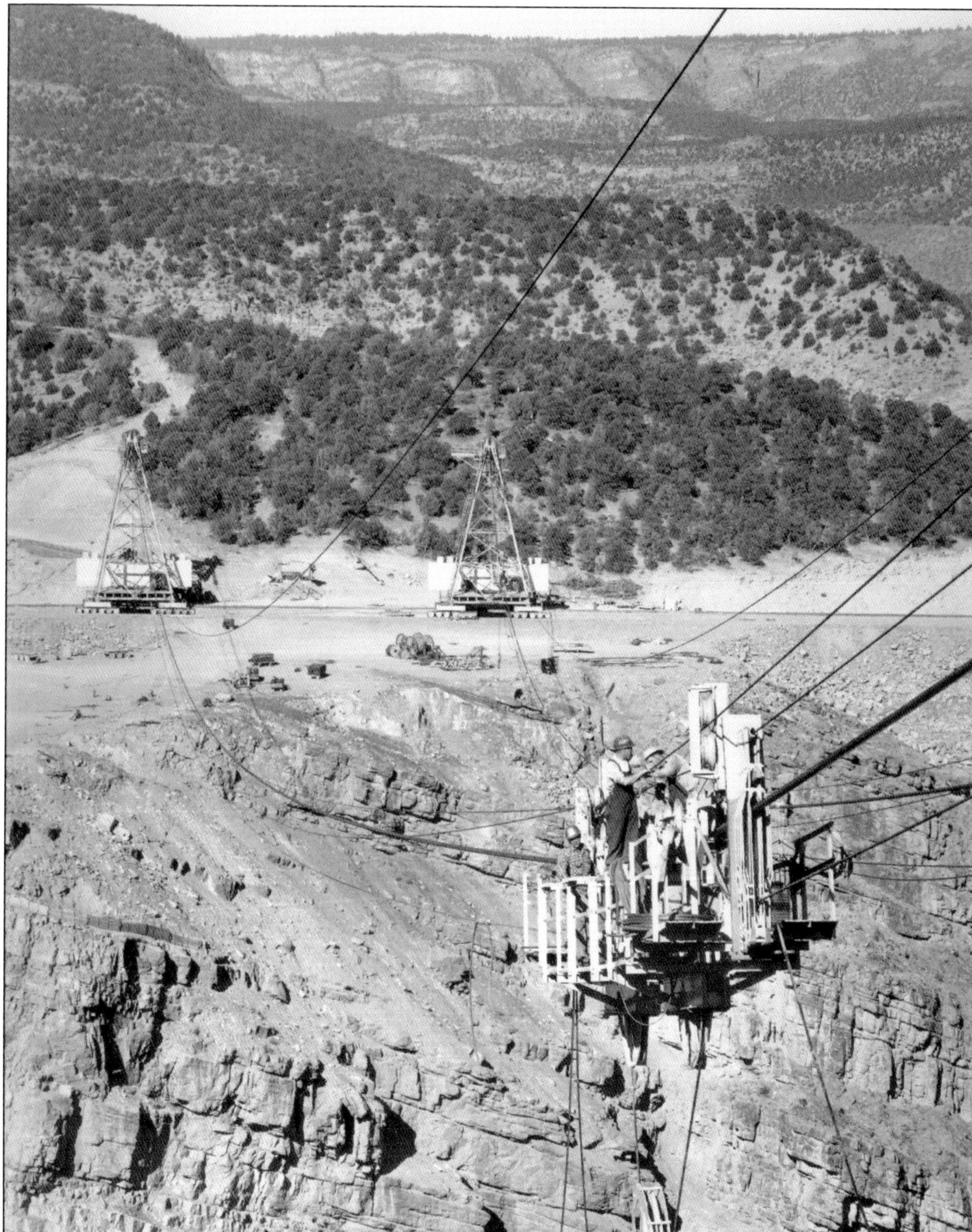

In August 1960, ironworkers ride the carriage as they finish work on the cableway system. The cables extend from a fixed headtower on the right abutment to two tailtowers on the left abutment of the dam. The movable tailtowers are 180 and 185 feet in height. The cables connecting them to the headtower are 1,900 feet in length. When concrete placing begins, it will require only three minutes to complete a cycle of delivering the concrete in eight-cubic-yard buckets to the bottom. The cable is suspended 500 feet above the Green River. (Courtesy of BOR.)

On October 27, 1960, workers of Selby Drilling Company are shown operating a diamond drill, preparing holes for a 1 1/2-inch anchor bar in the spillway intake area of the left abutment of the dam. The anchor bar to be installed will pin the surface rock to massive subsurface rock. (Courtesy of BOR.)

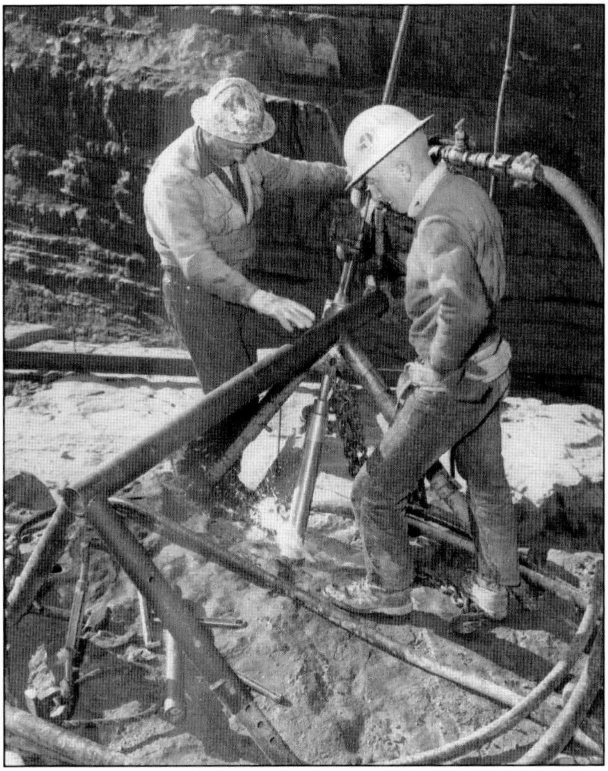

Arch Dam Constructors, contractors for the Flaming Gorge Dam, begin drilling at the downstream portal of the spillway tunnel. A black outline is seen on the rock ledge where the tunnel must be created by the portable drilling machines. This photograph was taken on August 24, 1960. (Courtesy of BOR.)

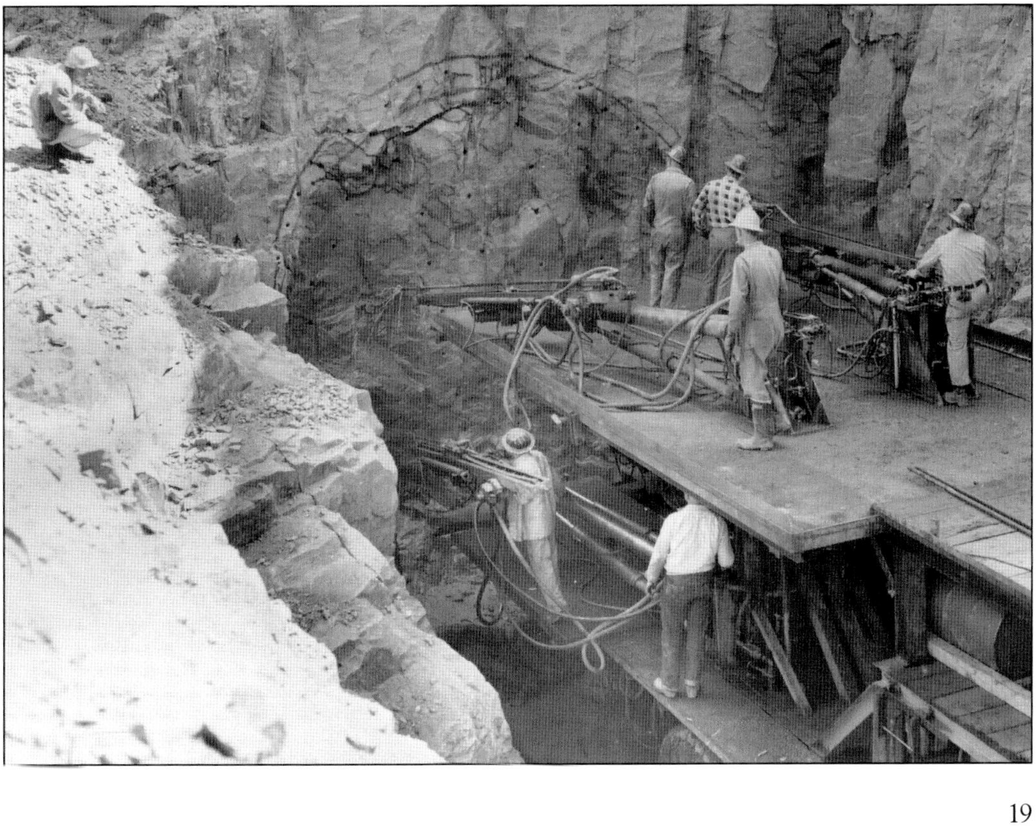

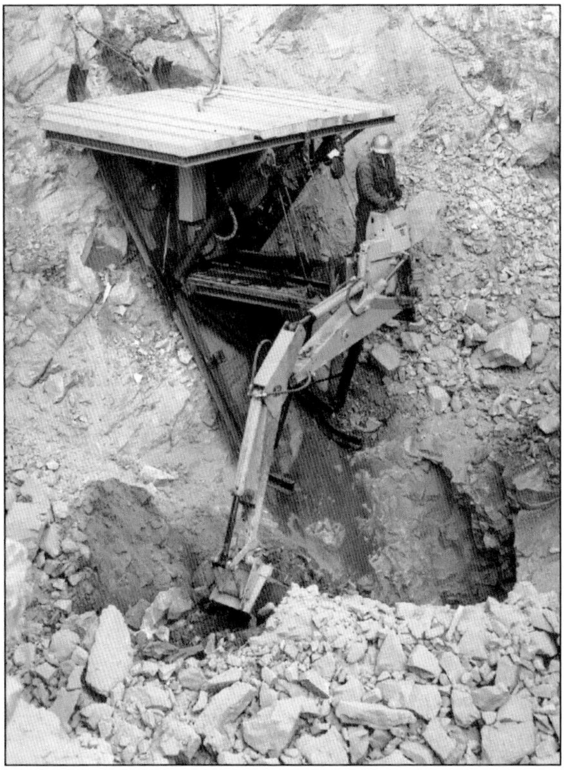

This mucking machine was created by Arch Dam Constructors for use in excavating blasted rock from the spillway tunnel. The hole below the machine operator is the six-foot-by-eight-foot pilot raise, driven from the bottom of the tunnel upward. The blasted material was mucked into the tunnel and then picked up at the lower end of the tunnel and disposed of. The sled and platform that form a part of the mucking device was lowered into the tunnel as the heading advanced. This photograph was taken on January 30, 1961. (Courtesy of BOR.)

The spillway tunnel is inclined on a 55-degree angle horizontally at the inlet end. The tunnel will be around 700 feet long when finished and vary in diameter, from 26.5 feet at the intake end to 18 feet at the outlet. It will be lined in concrete. (Courtesy of BOR.)

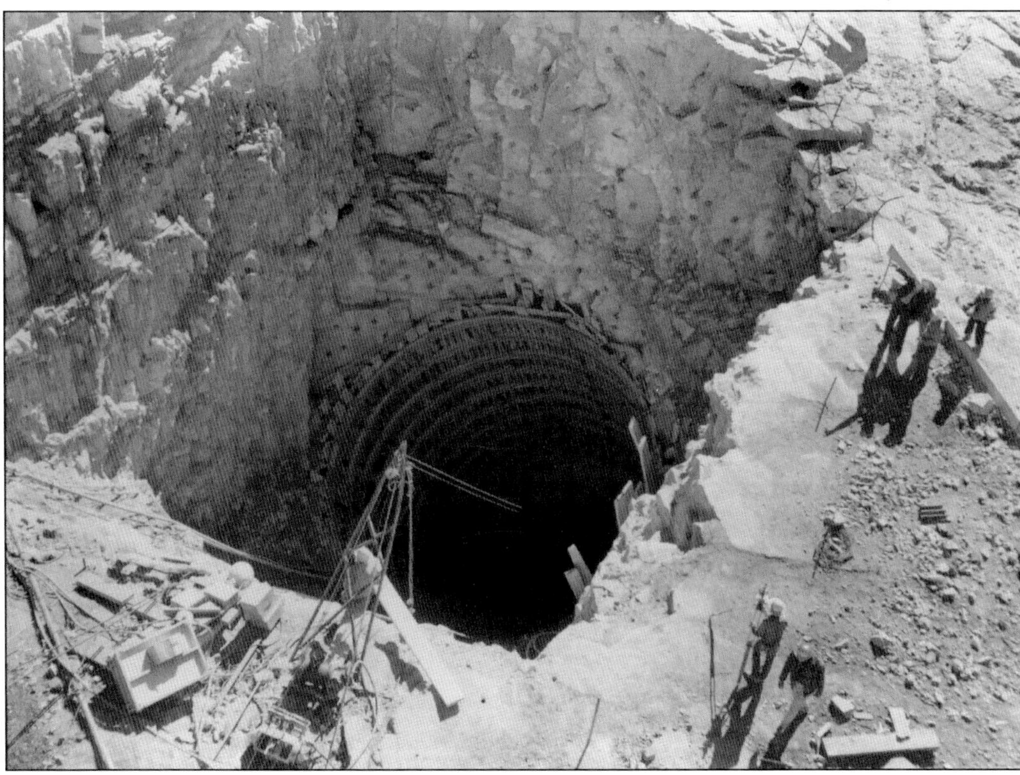

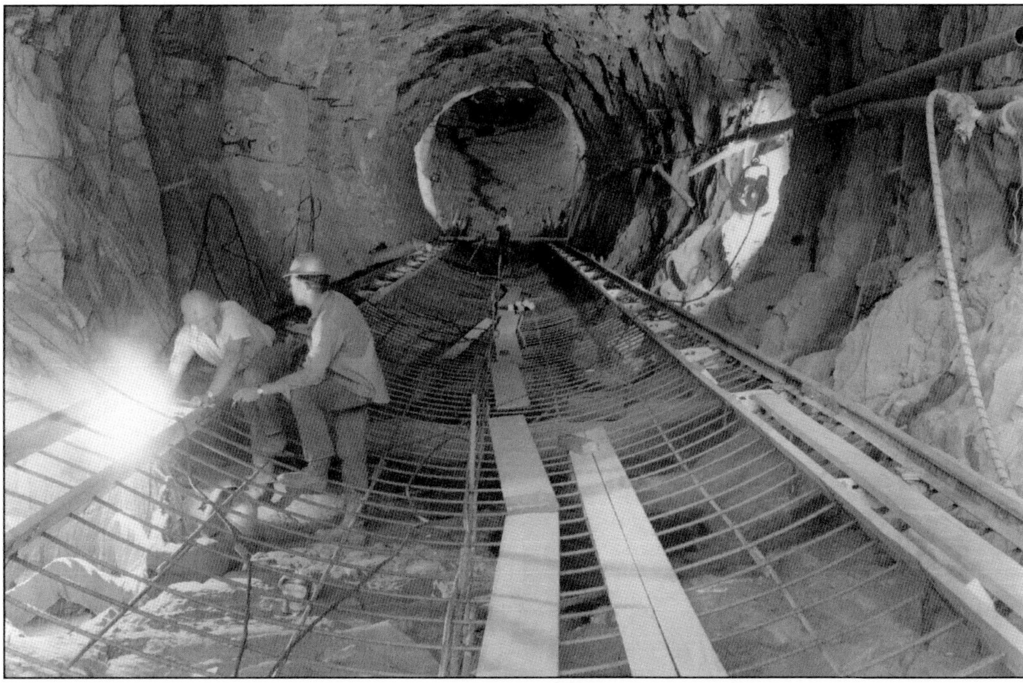

Tunnel workers weld reinforcement steel mat for the placement of concrete in the spillway tunnel of the dam on June 27, 1961. Pat Campbell of Oreville, California, is underneath doing the welding. On top are D.R. Merrit (left) of Afton, Wyoming, and Darrel Walker of Salt Lake City. In the background, using a survey instrument, is Lee Harmston of Roosevelt, Utah. (Courtesy of BOR.)

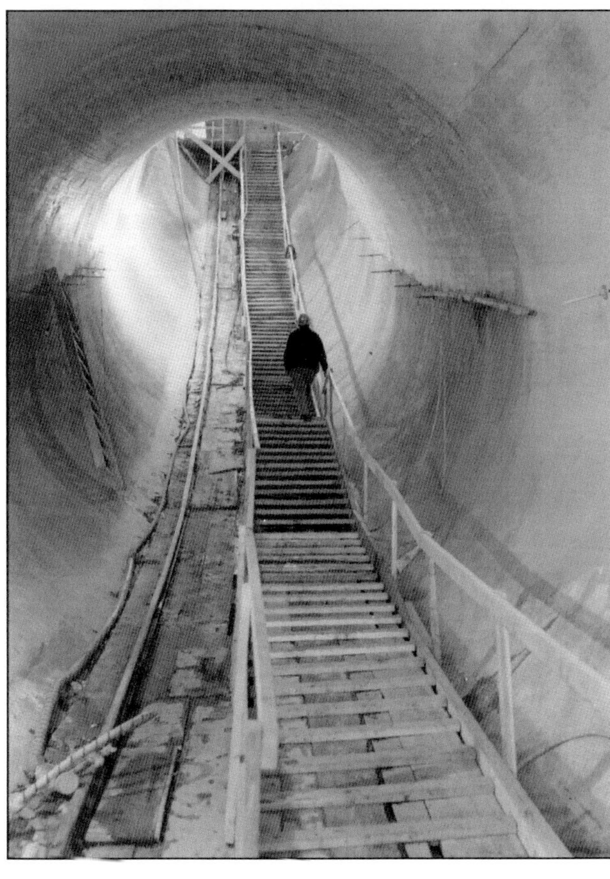

A man is seen climbing the steps inside the spillway tunnel on March 29, 1962. The view is looking upstream along the 55-degree incline of the tunnel. (Courtesy of BOR.)

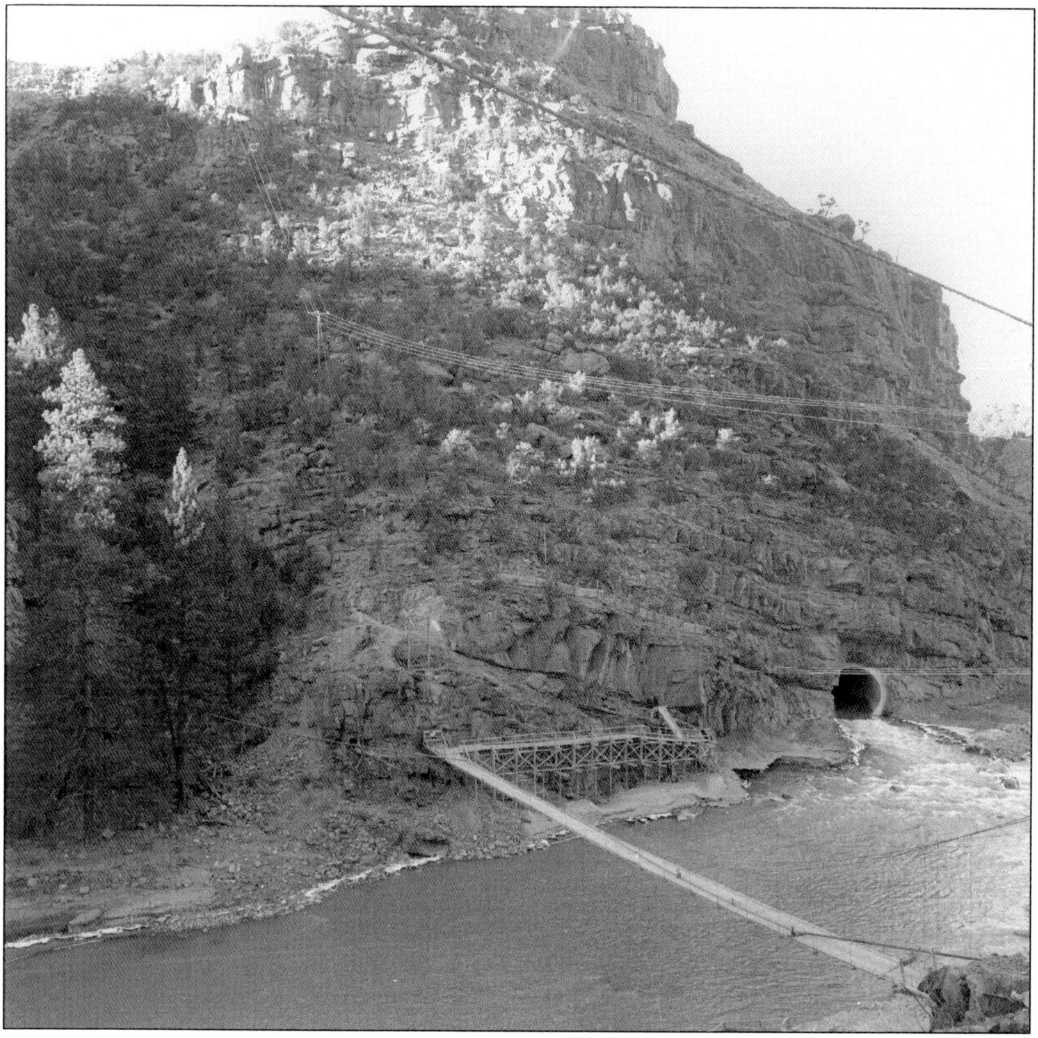

A catwalk, built across the Green River below the diversion tunnel outlet, allowed construction workers to cross the river. A steady flow of water is seen running from the outlet below the Flaming Gorge Dam construction in December 1959. Above the tunnel is the cofferdam, which holds back the water from the construction site; it is 260 feet long, 80 feet through the base, and 23 feet above the water level. Another cofferdam was built upstream from the dam construction. (Courtesy of UCRHC.)

Two

A New Town Is Born

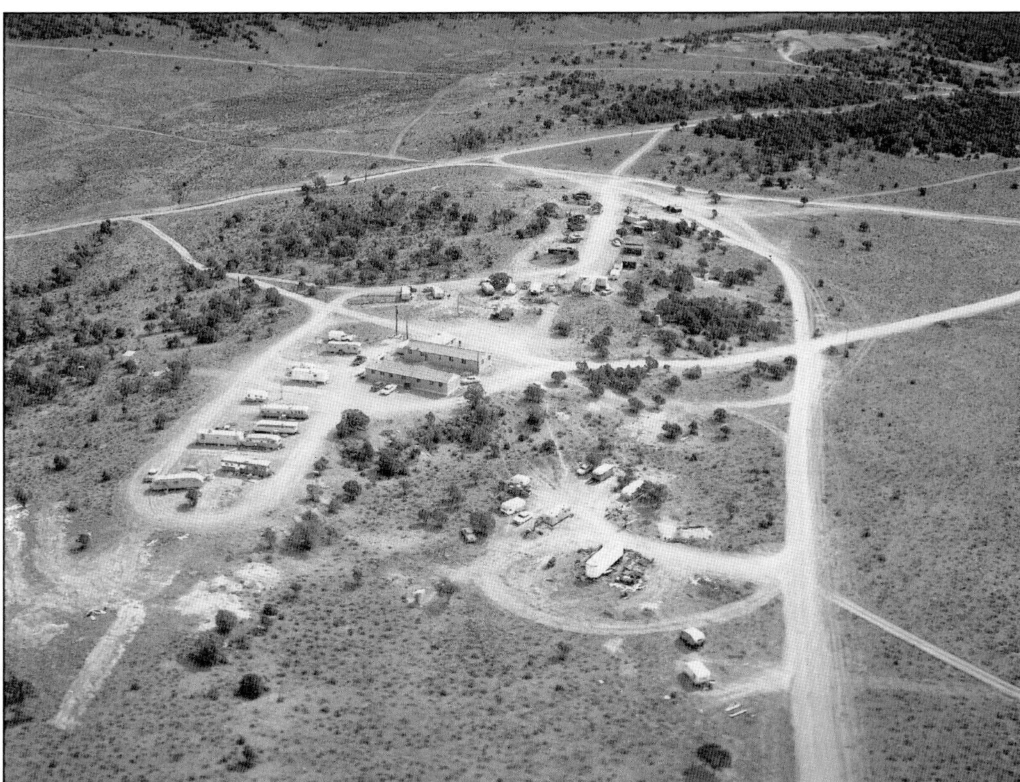

In this aerial view of Witt Construction Company's camp at the Flaming Gorge community, the two larger buildings are a bunkhouse and a mess hall for employees. The Witt Company arrived ahead of the dam construction crews to clear land that would eventually be covered by the reservoir. This work included removing all buildings, bridges, cattle guards, fencing, and other improvements from the reservoir site below an elevation of 6,040 feet. (Courtesy of BOR.)

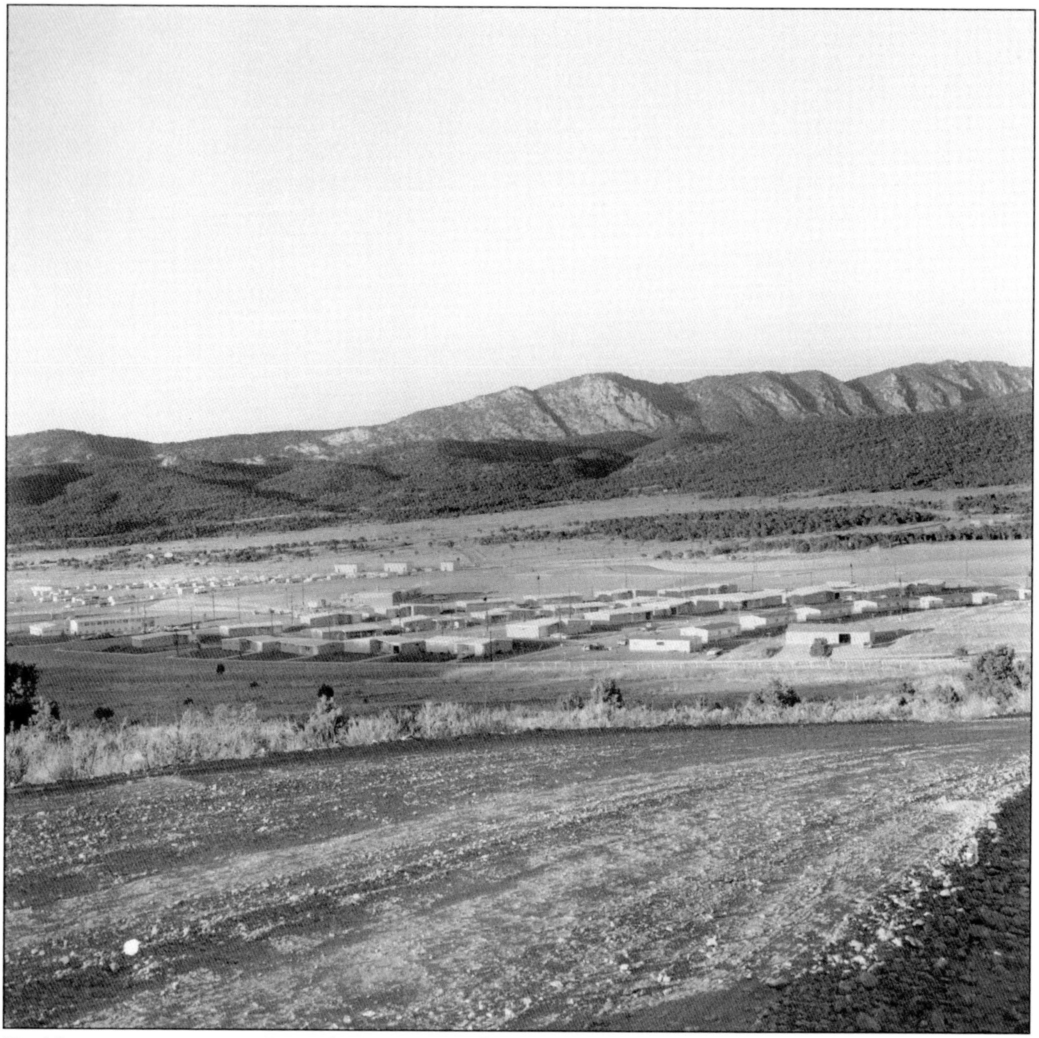

Utah's newest town is taking shape among the sagebrush and wilderness of Dutch John Flat, a desolate bench on the river's north side. Dutch John was named after John Honselena, an early settler from Sheiswig, Germany. To the people of the time, he sounded like a "Dutchman," hence the name. Of the 25 transit homes to be erected, 23 are in place fronting South Boulevard. This booming town will include warehouses, a service station, a modern hospital, a post office, a modern sewage disposal system, a water-treatment plant, several water storage tanks, and an airport. A new school (left center) is being erected by the State of Utah. Dutch John became not only the newest but also the largest town in Daggett County, although it never became an incorporated town. (Courtesy of UCRHC.)

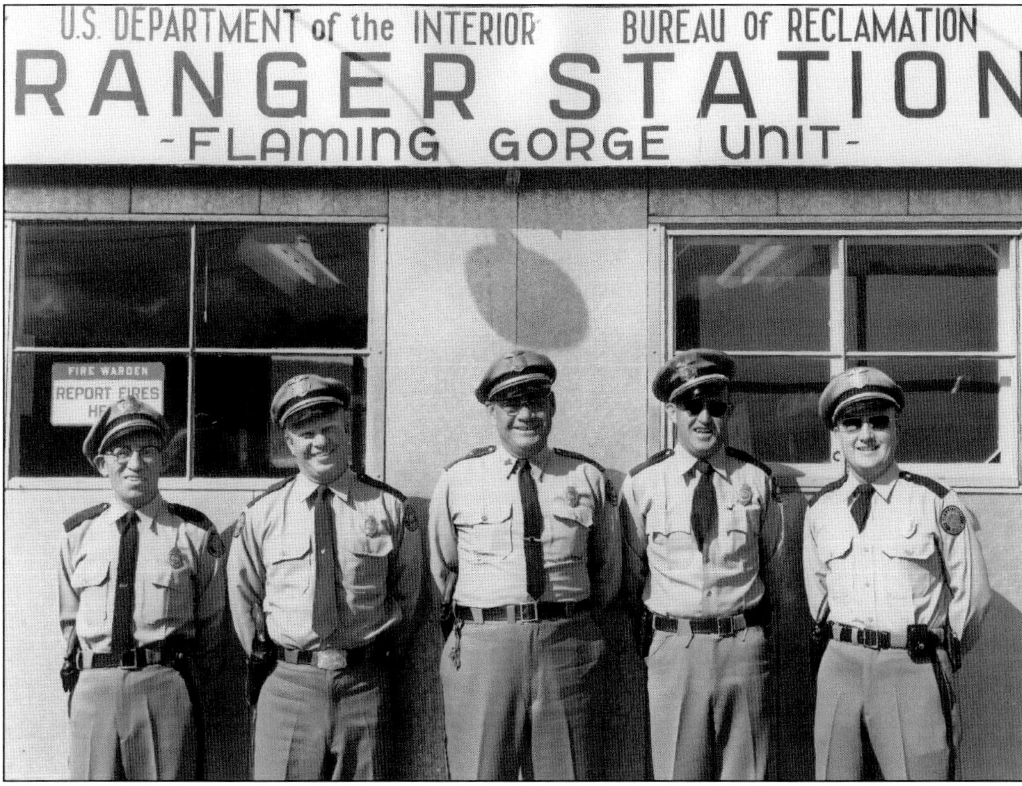

During the height of construction on Flaming Gorge Dam, additional law enforcement was needed in the Dutch John area. Pictured here are members of the Bureau of Reclamation Police Force at Dutch John. From left to right are Capt. William E. Fletcher, Elmer W. Greathouse, Chief Orval A. Martin, Gene Campbell, and Clarence Chidester. These men took care of law enforcement problems, which were relatively few due to the remoteness of the area. (Courtesy of BOR.)

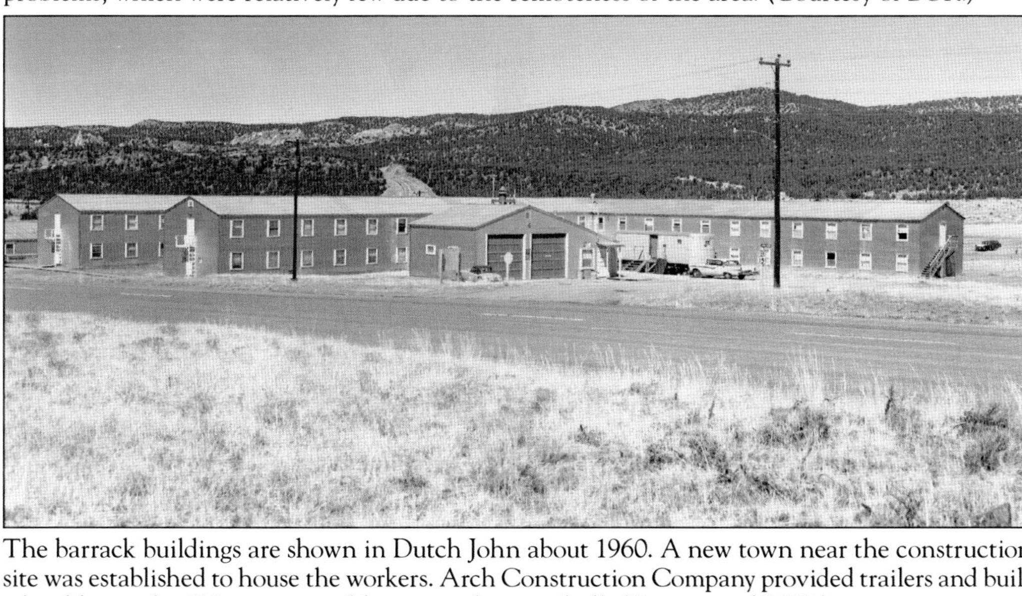

The barrack buildings are shown in Dutch John about 1960. A new town near the construction site was established to house the workers. Arch Construction Company provided trailers and built a bunkhouse for 300 men, a cookhouse, and a mess hall. (Courtesy of BOR.)

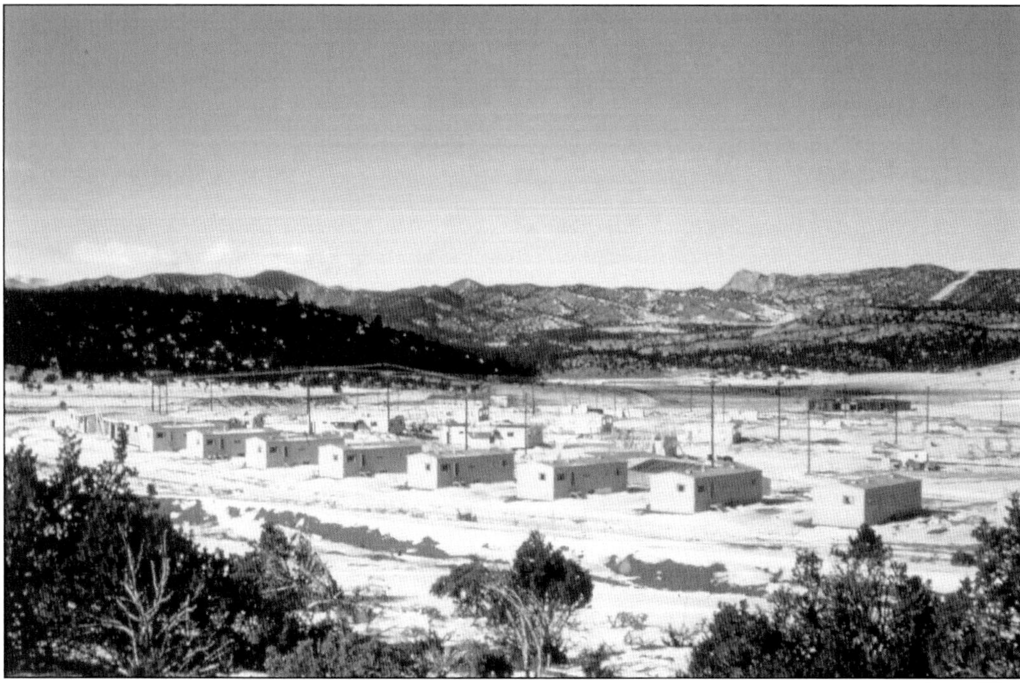

To house and serve the government workers, eight transit homes and 25 trailers were installed in 1957. A large 60-foot-long trailer served as a hospital; for a time, it had the only bath in town. Other trailers served as a fire station and a post office. (Courtesy of BOR.)

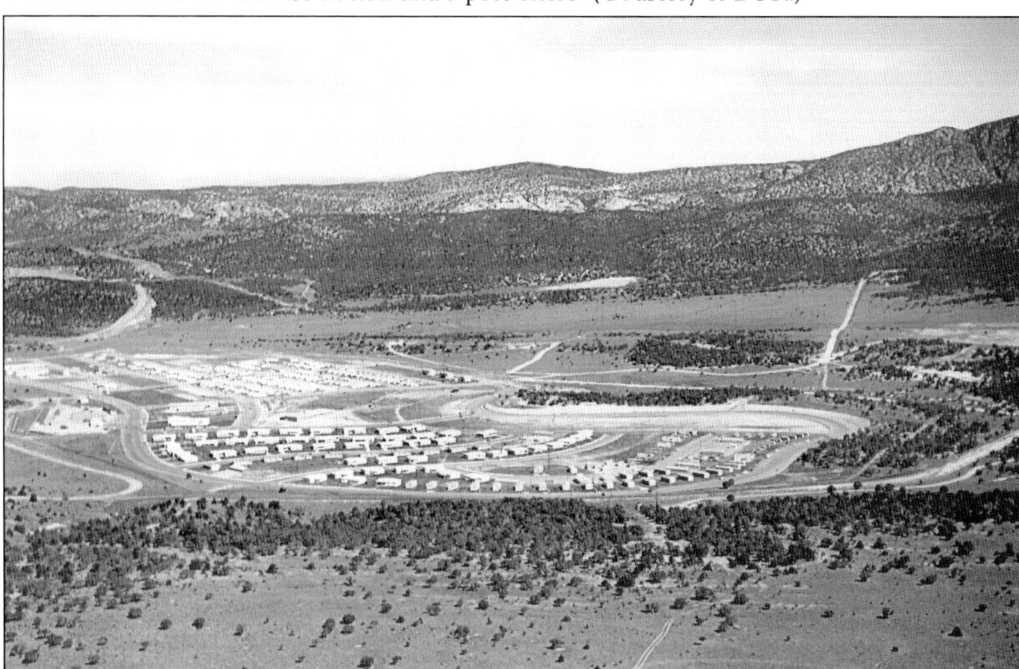

At the peak of construction activity, Dutch John reportedly had a population of over 3,500 workers. Work commenced on permanent buildings, including homes, a school, a 10-bed hospital, and a firehouse. (Courtesy of BOR.)

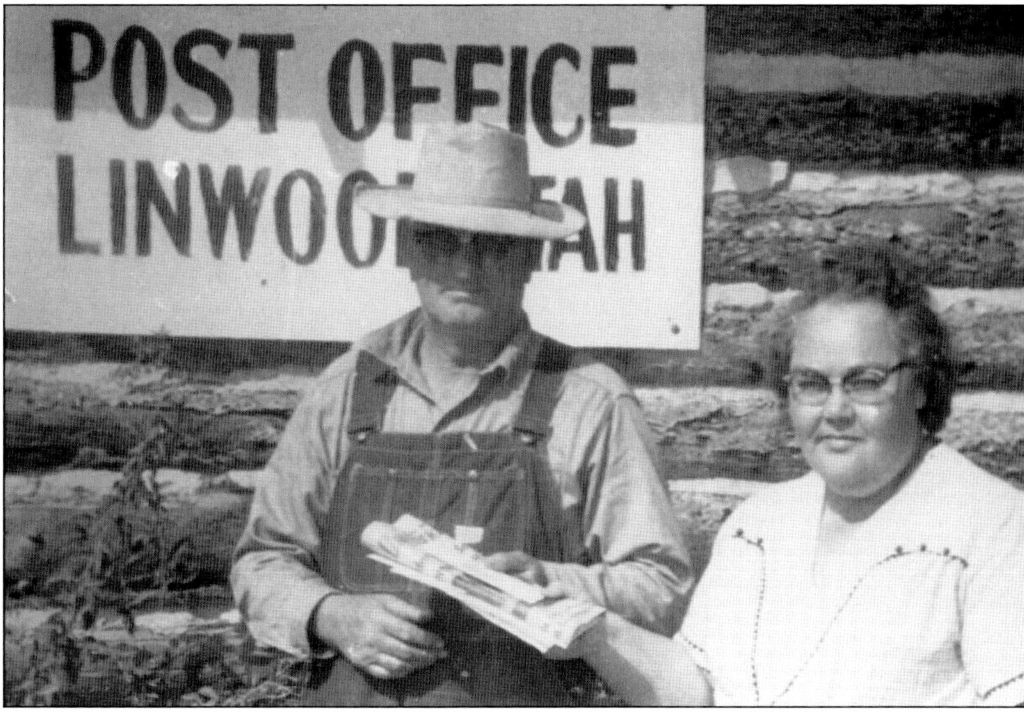

In the above photograph from September 1958, postmaster Eleen Williams hands Tim Potter his mail from the last mailbag delivered to the Linwood Post Office. When Flaming Gorge Dam was built, Williams moved with the post office to Dutch John. She said, "For a while, no one knew for sure what the name of the new town was." Letters to the new post office were variously addressed to "Long John," "Uncle John," "Brother John," and "Dutch Jack." The building in the photograph below, from around 1960, housed the Dutch John Family Store, the post office (on the left end of the building), and a snack bar with a drive-up window. Williams said that between 1960 and 1964 she had to hire six clerks to handle the 3,000 patrons in Dutch John. (Above, courtesy of Pauline Baker; below, BOR.)

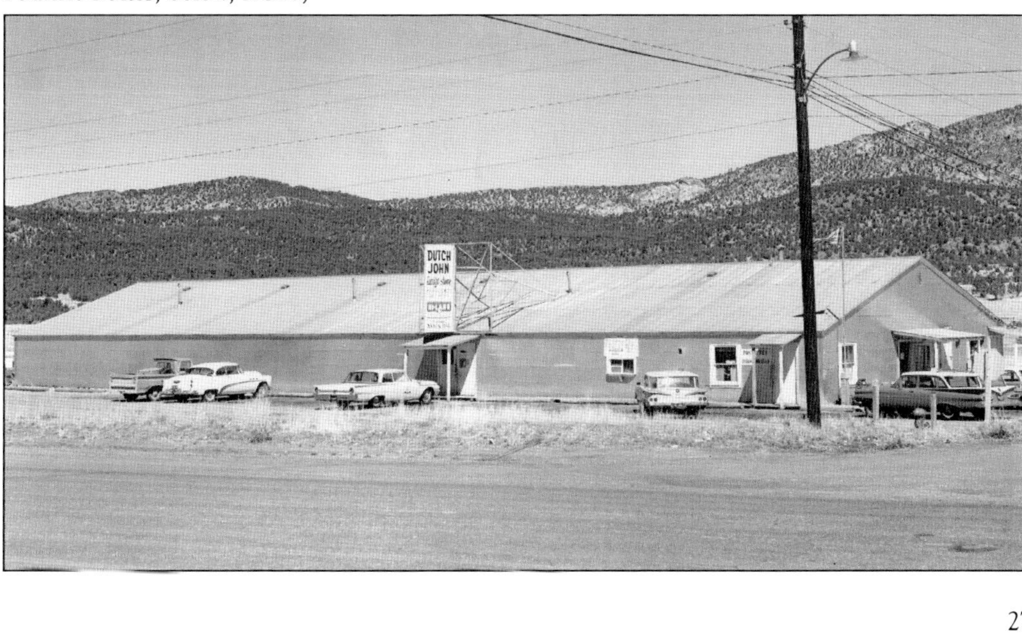

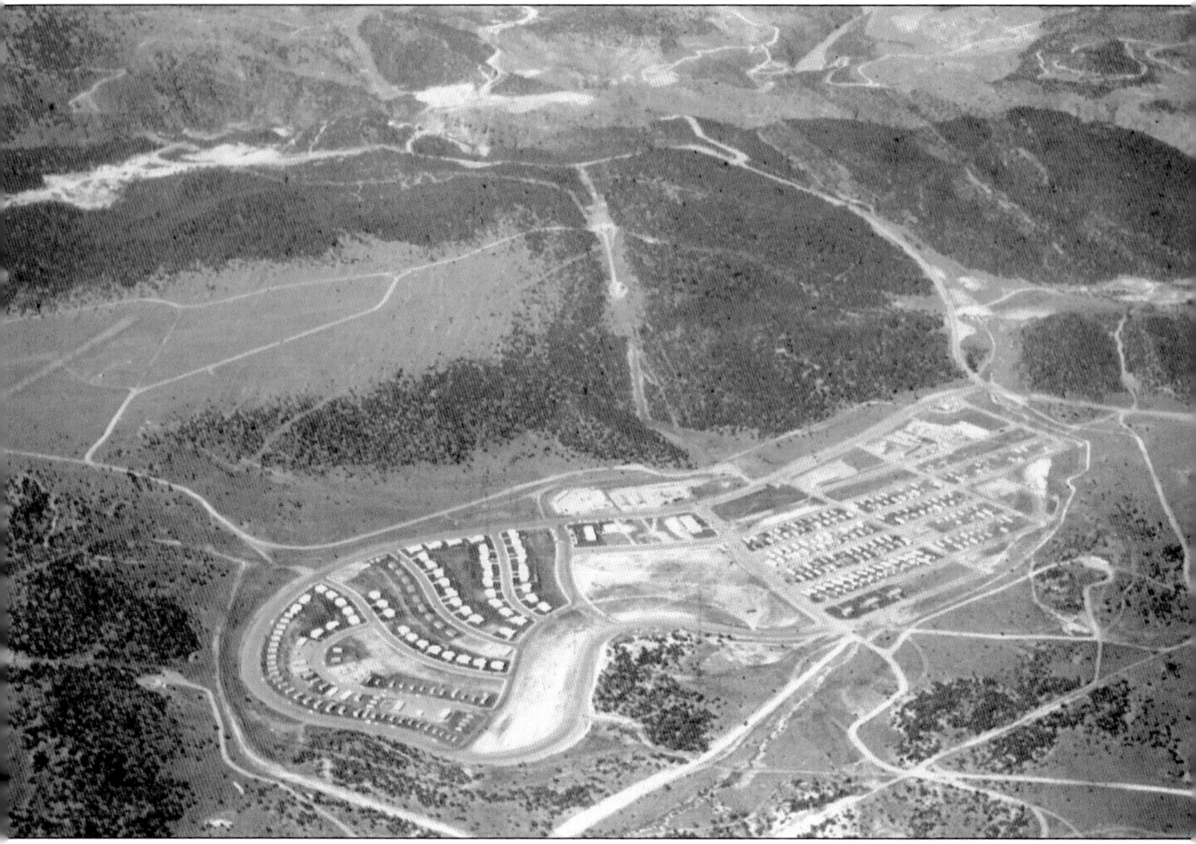

When work on the dam was completed, all of the contractor's facilities were removed except for one building, which was remodeled to become the LDS church. The Bureau of Reclamation moved offices to the dam, which is only three miles from the town, and the Forest Service took over the building in Dutch John. The store and service station have remained; an airstrip was added in 1968. As of this printing, the community's population varies from nearly 250 in the summer to about 150 during the winter months. Dutch John has an airport with a paved landing strip 6,000 feet in length. Building and completing the dam was a bittersweet event for many residents of Daggett County. (Courtesy of BOR.)

Three

LINWOOD
"THE TOWN THAT BURNED THEN DROWNED"

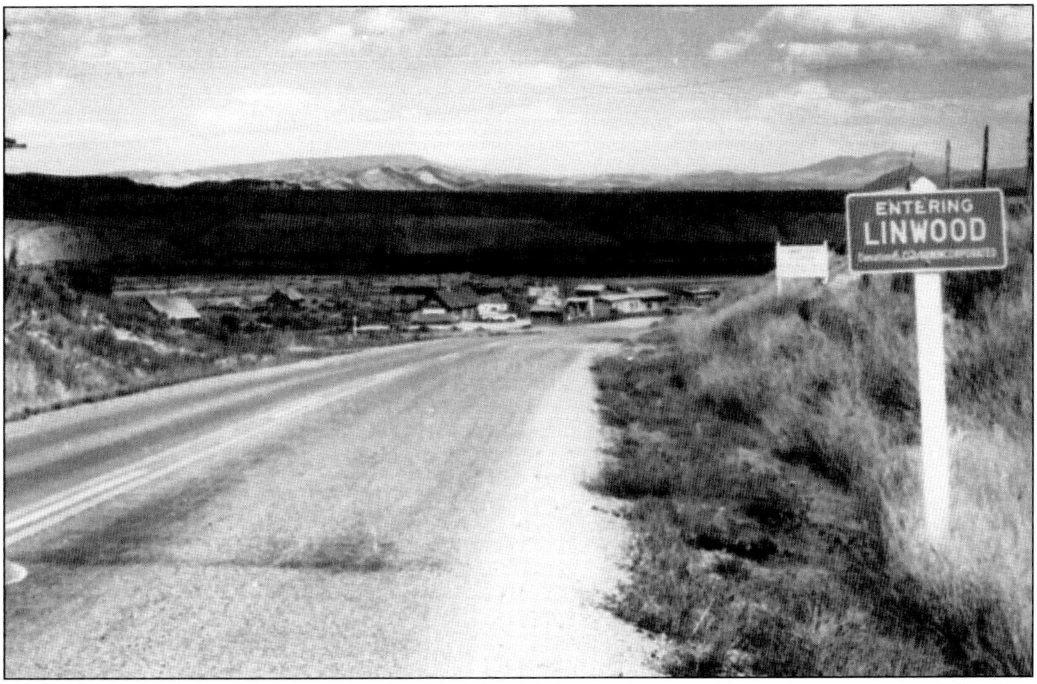

Linwood was Daggett County's first town. In 1836, Jack Robertson, a mountain man and the first resident, constructed a cabin on lower Henry's Fork. It served as a trading post for trappers and Native Americans. By the 1880s, the settlers on Henry's Fork created a community. Linwood, frequently visited by Butch Cassidy and the Wild Bunch among others, attained prominence. It became a favorite supply point for the Outlaw Trail. (Courtesy of Daggett County.)

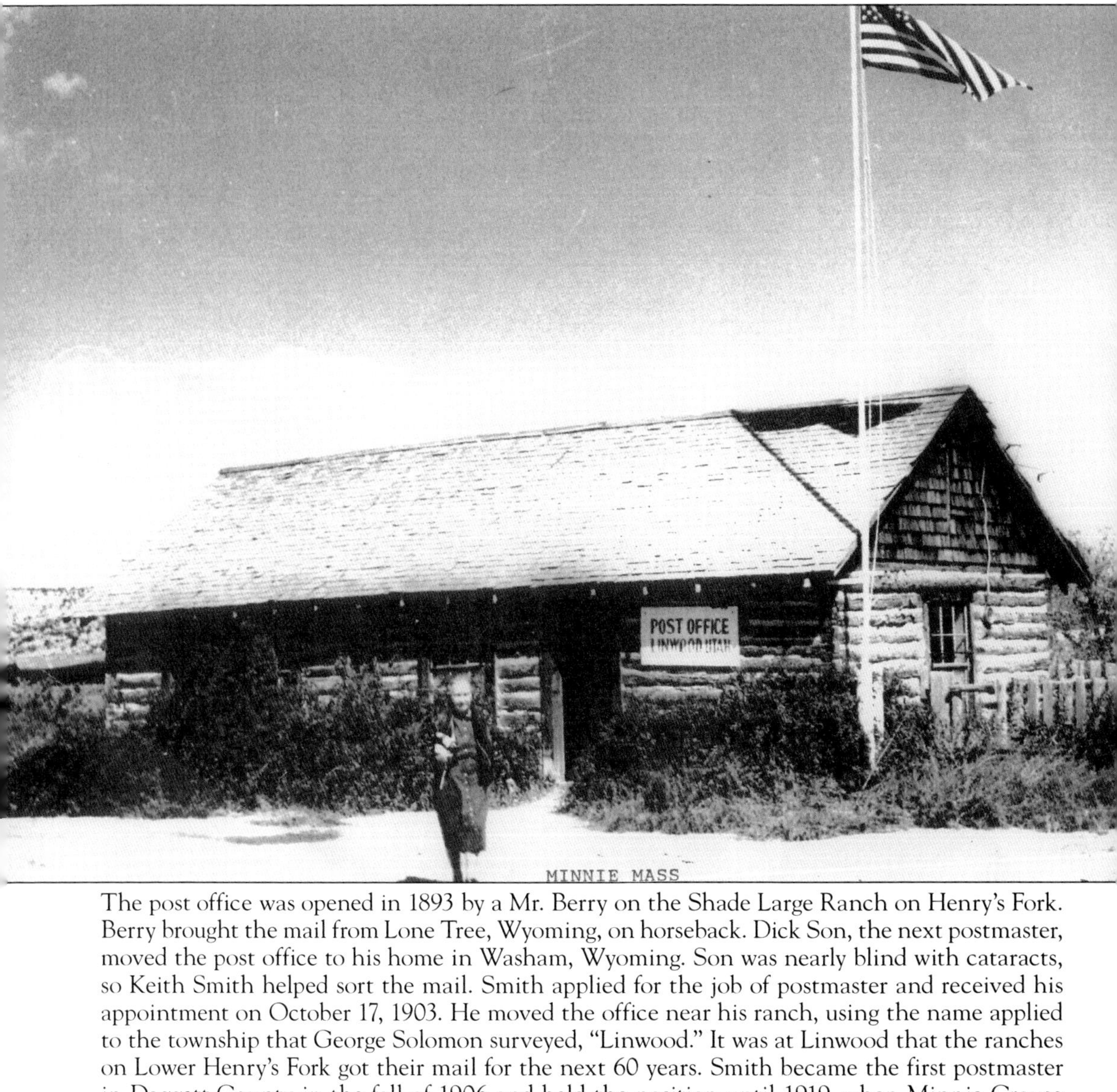

The post office was opened in 1893 by a Mr. Berry on the Shade Large Ranch on Henry's Fork. Berry brought the mail from Lone Tree, Wyoming, on horseback. Dick Son, the next postmaster, moved the post office to his home in Washam, Wyoming. Son was nearly blind with cataracts, so Keith Smith helped sort the mail. Smith applied for the job of postmaster and received his appointment on October 17, 1903. He moved the office near his ranch, using the name applied to the township that George Solomon surveyed, "Linwood." It was at Linwood that the ranches on Lower Henry's Fork got their mail for the next 60 years. Smith became the first postmaster in Daggett County in the fall of 1906 and held the position until 1919, when Minnie Crouse Rasmussen, shown standing in front of the Linwood Post Office, began her 33-year tenure as postmaster. Eleen Williams headed up the post office in Linwood until she moved with it to Dutch John in September 1958. (Courtesy of Daggett County.)

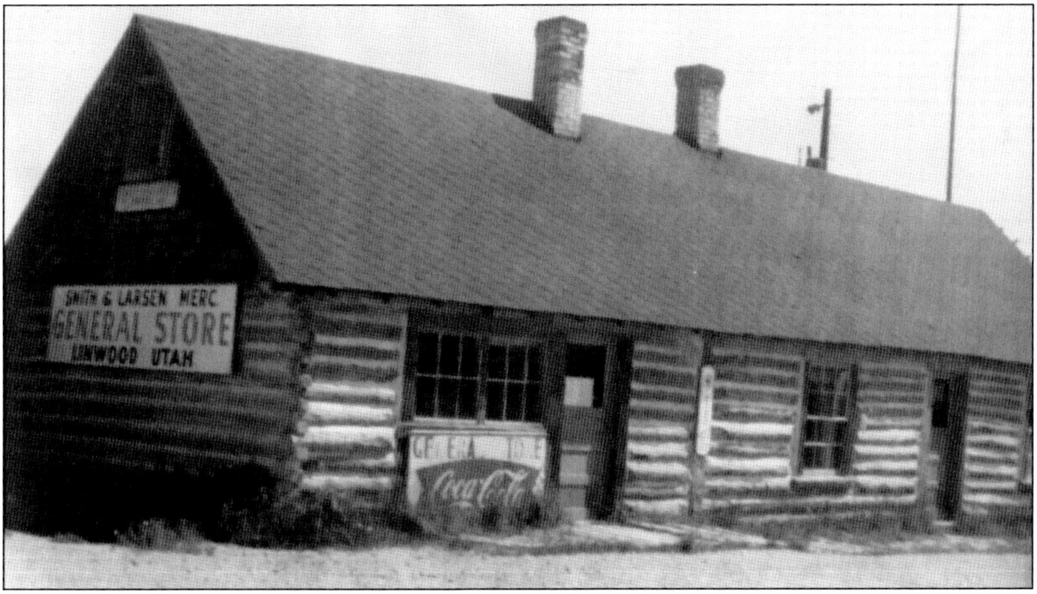

A month after the post office came to Linwood, Keith Smith and his neighbor Marius Larsen opened the Smith & Larsen Mercantile Company store at the opposite end of the same building. It became a focal point in the community. A boom in the sheep business stimulated development in the Linwood area. (Courtesy of Daggett County.)

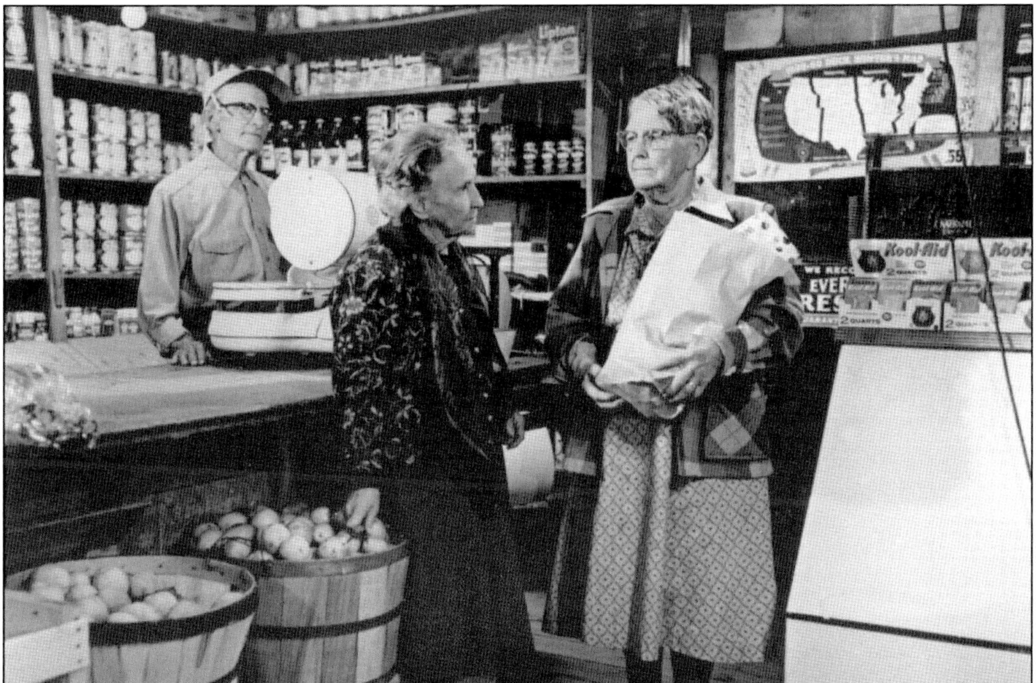

Shown here are George Rasmussen (left), Minnie Rasmussen (center), and Mabel Adamson. The Smith & Larson Mercantile Company store had been run by George since 1909. Everyone from miles around had fond memories of meeting friends there. George took great pride in serving his customers and kept the stock in the small building in order. (Courtesy of Daggett County.)

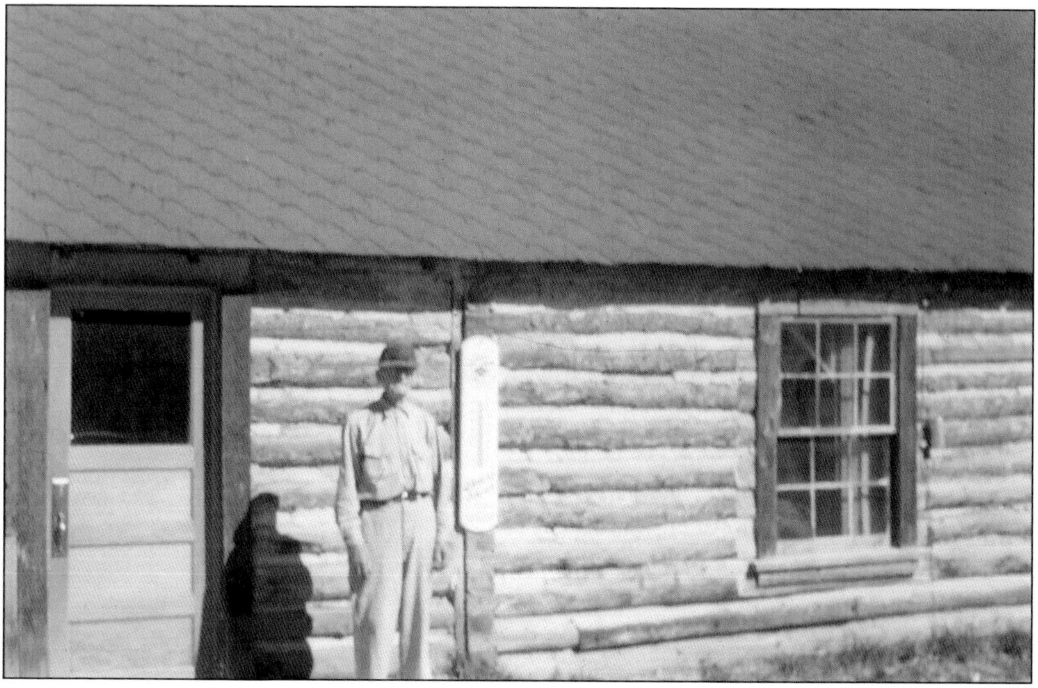

George Rasmussen locked the front door of the general store. He stood a moment as memories of more than half a century flooded his mind. He sighed, shrugged his shoulders, and walked slowly away. Turning the key in the Linwood store marked the end of 52 years of service and sounded the death knell to a community that soon would cease to exist. (Courtesy of Pauline Baker.)

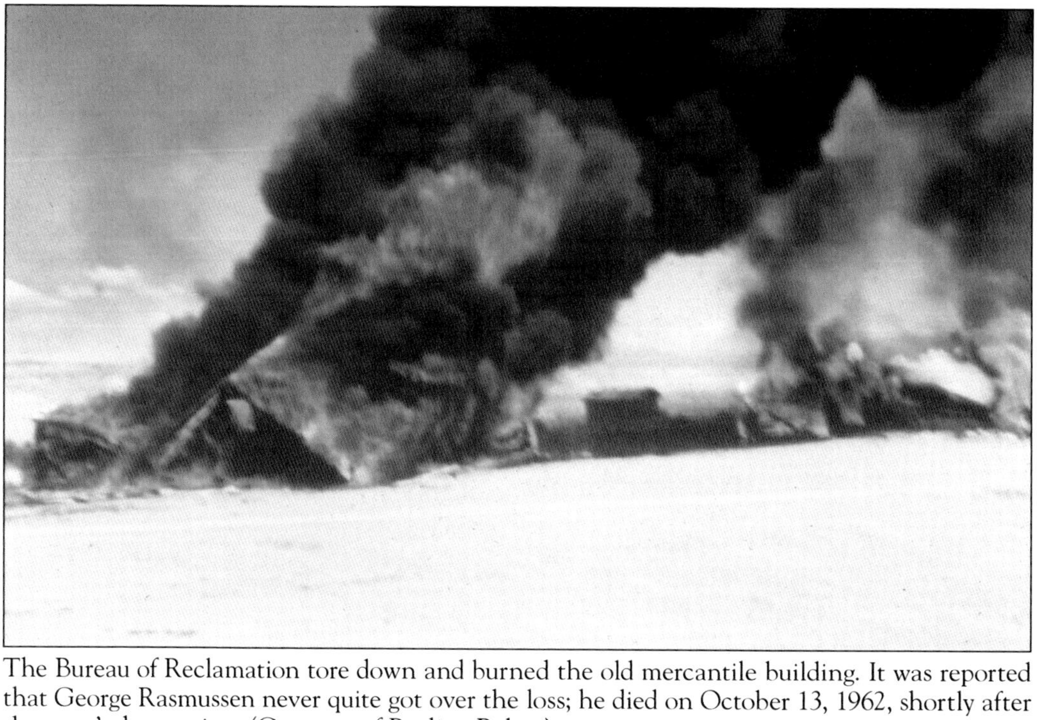

The Bureau of Reclamation tore down and burned the old mercantile building. It was reported that George Rasmussen never quite got over the loss; he died on October 13, 1962, shortly after the store's destruction. (Courtesy of Pauline Baker.)

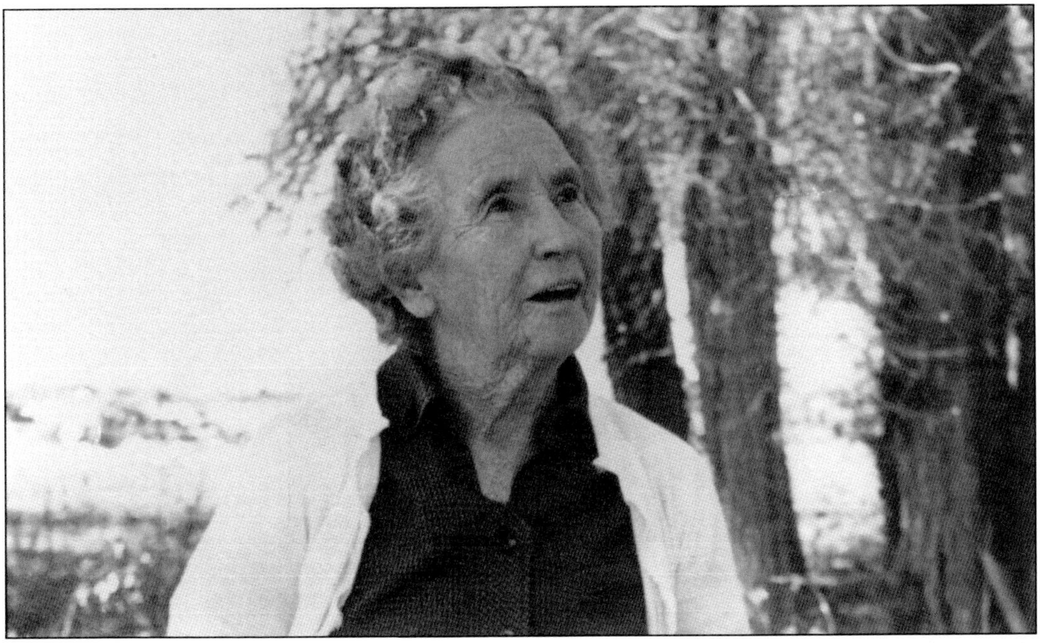

Minnie Crouse Rasmussen was born on June 5, 1882, near Bridgeport in Brown's Park. In 1917, she became the postmaster of Linwood and held that position until 1952. In 1924, she married George Rasmussen, owner and operator of the Linwood Merc (the Smith & Larsen Mercantile). Minnie lived in Linwood until 1963, when the quiet little town was cleared for the already rising Flaming Gorge Lake. On the afternoon of July 26, 1963, government officials stopped to visit Minnie to inform her that she must abandon her house (below), and she invited them in for tea. As the officials drove away, Minnie put a torch to her home rather than turn it over to the Bureau of Reclamation. For many Daggett County residents, Minnie's burning home became a symbol of defiance to bureaucratic authority and the many changes brought about by Flaming Gorge Dam. (Above, courtesy of Daggett County; below, Pauline Baker.)

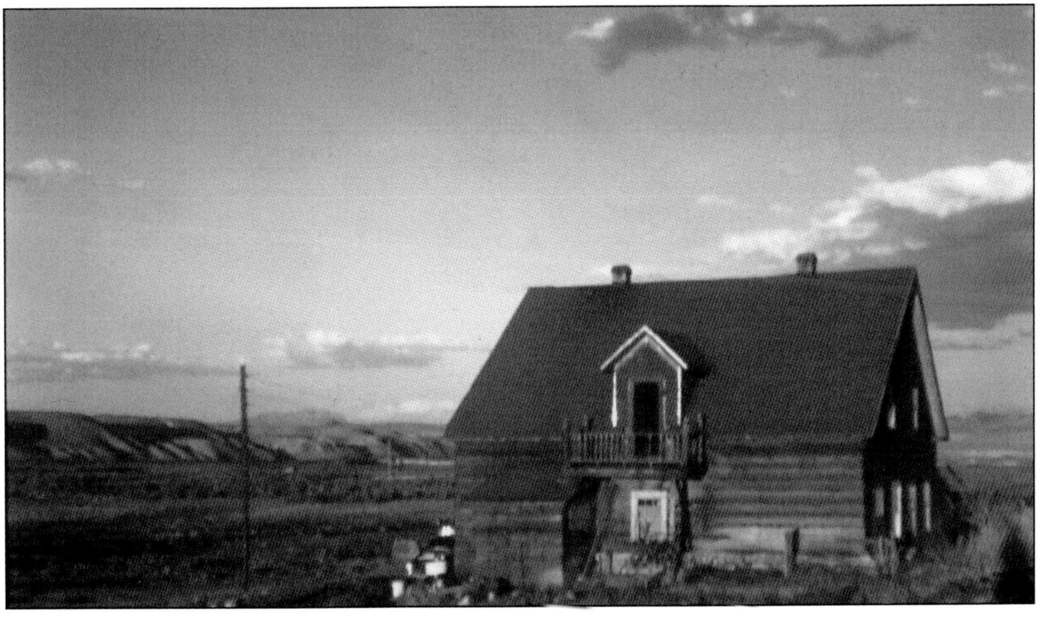

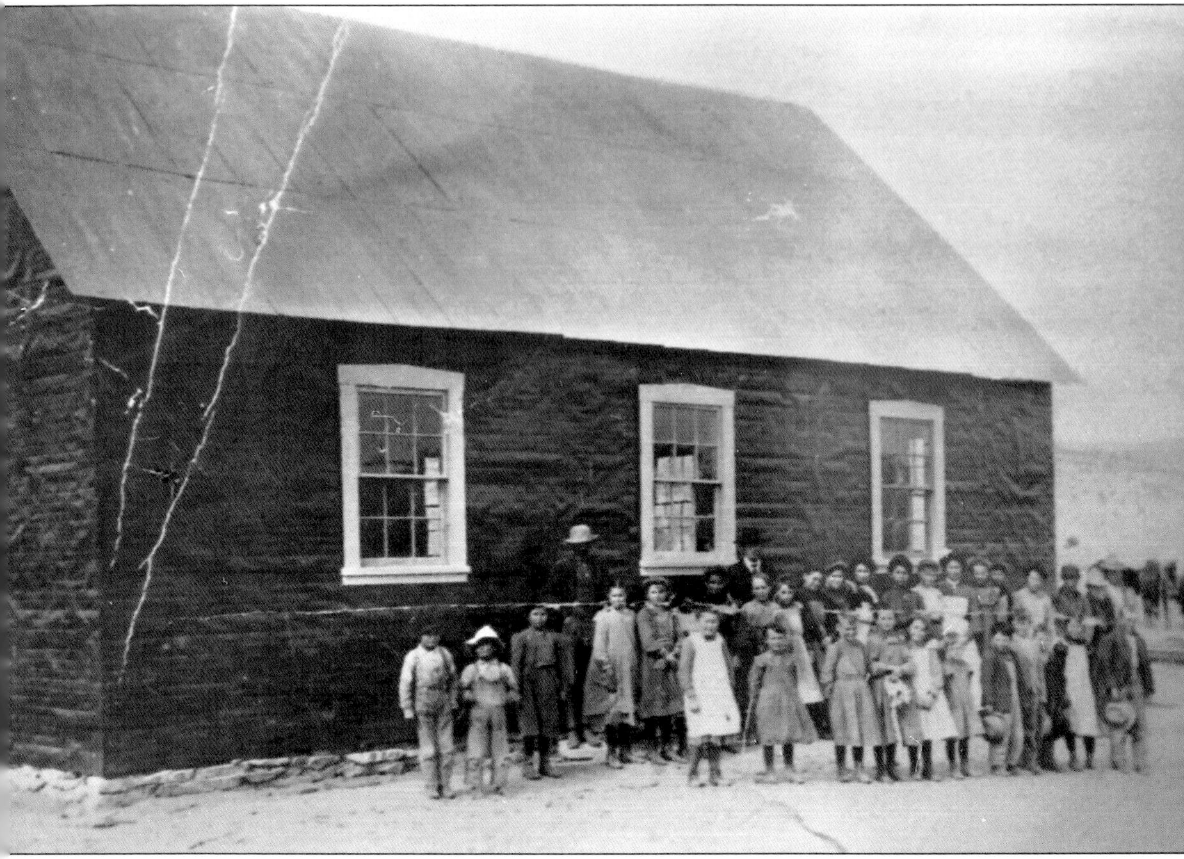

Daggett County, Utah, and Sweetwater County, Wyoming, share an interstate oddity: a one-room schoolhouse that stands on the state line, which runs through the middle of the room. The Red School House was so arranged that one teacher could teach the children in the two states. The children from Utah sat on the south side of the aisle and those from Wyoming on the north side so that all attended school in their home state. Mr. Pinkney was the first teacher at the school. Utah law forbade corporal punishment, while Wyoming stated that a whopping was all right. So, when a Utah student needed a bit of discipline, Pinkney just brought the student over to the Wyoming side and administered the proper number of whacks. The old school was moved to Manila to save it from the rising waters of the reservoir. (Courtesy of Mark Wilson.)

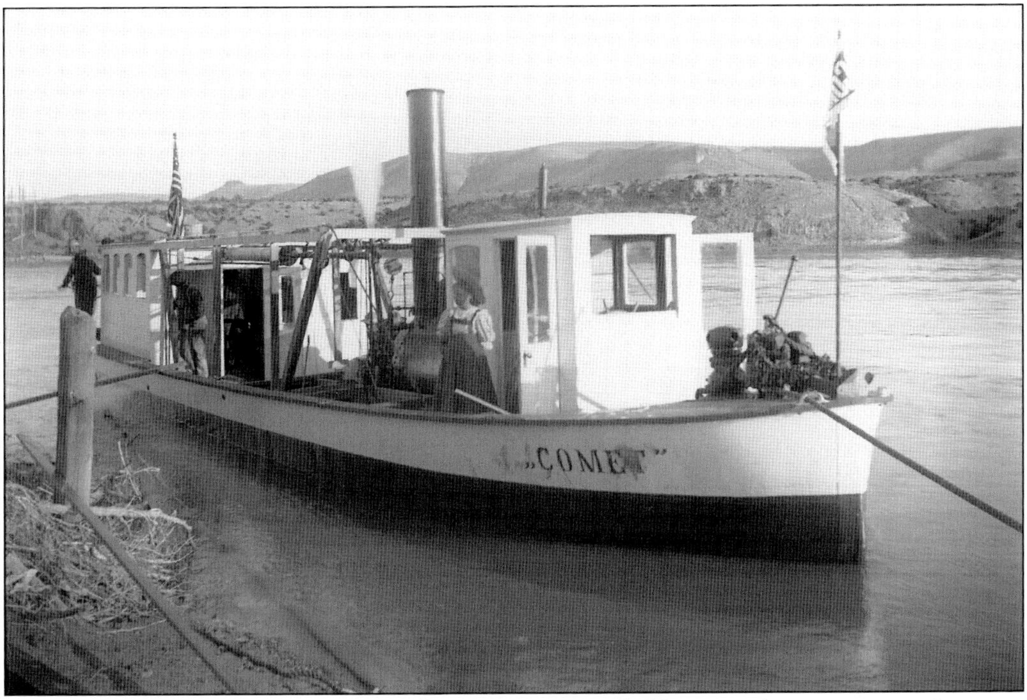

The steamboat *Comet*, built at a cost of $25,000, was intended for general transportation between Green River, Wyoming, and Linwood, Utah, carrying passengers and freight to various places along the river. The boat's maiden voyage, on July 7, 1908, took eight hours. Passengers enjoyed good entertainment, delicious meals, and beer. The return trip, however, was no party. Over a 30-hour period, the *Comet* ran aground and out of fuel many times. It made one more trip to Linwood but could not carry enough fuel for the round-trip. In 1908, it was evident that steam-boating on the Green River was impractical. The boat was stripped and eventually sank. (Courtesy of Sweetwater County History Museum.)

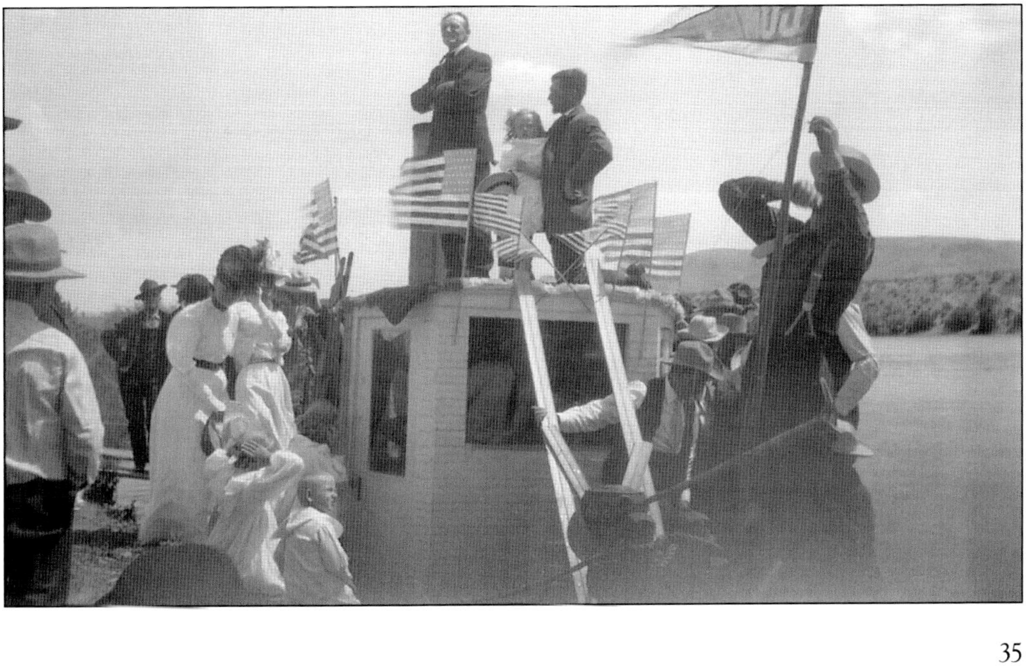

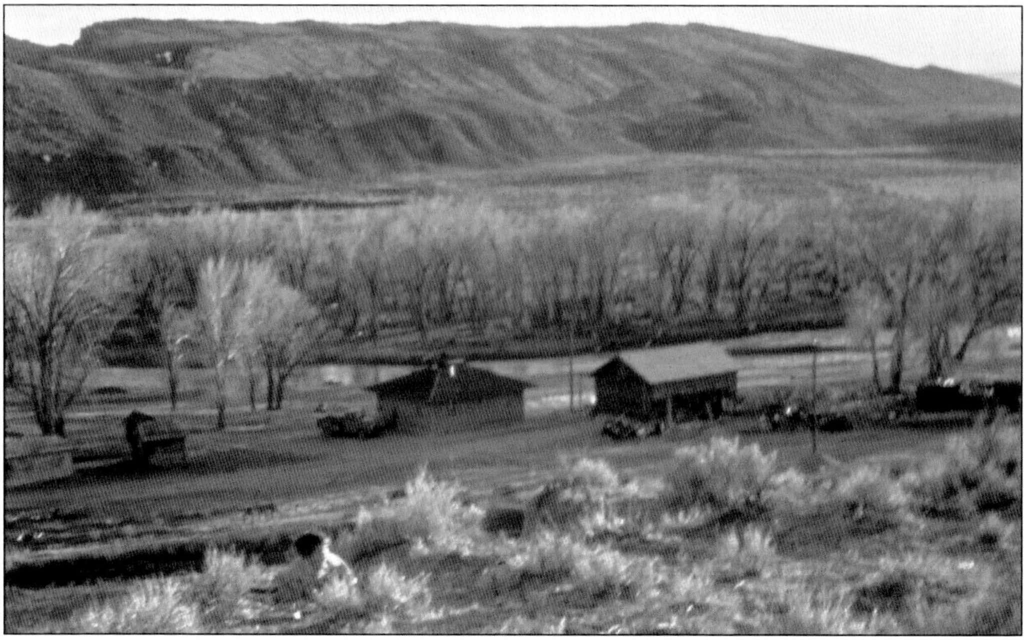

The completion of Flaming Gorge Dam was a bittersweet event for many in Daggett County. The community of Linwood had been doomed by the rising waters of the lake, and federal regulations specified that all reservoir sites below water level had to be stripped clean. This included the Williams Brothers Ranch at Henry's Fork, shown here before construction of the dam. (Courtesy of Pauline Baker.)

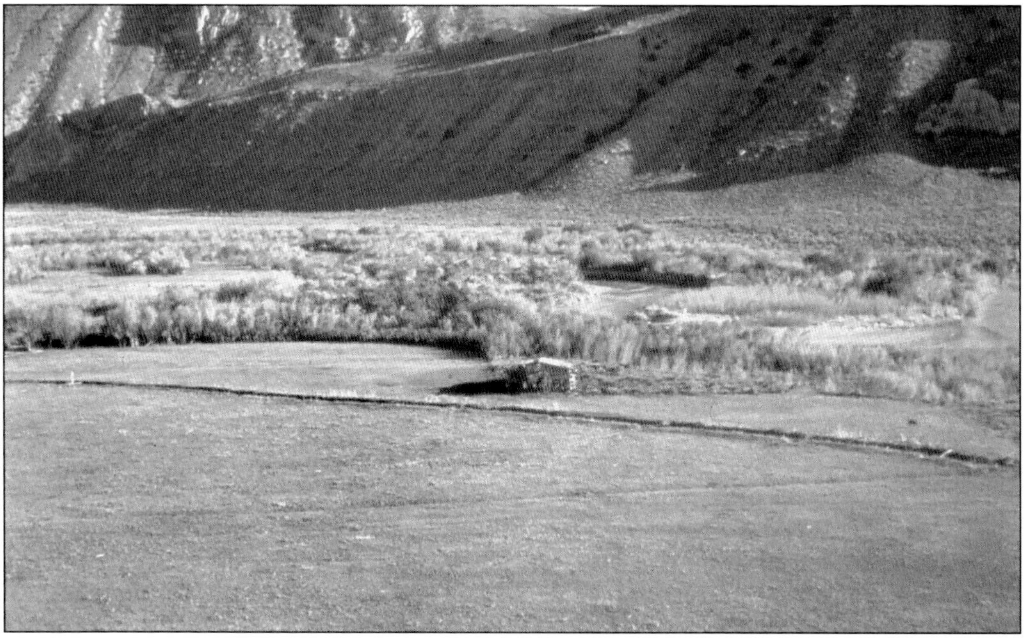

In 1958, the Williams brothers put up their last hay crop. They moved their home, originally part of the old Driskell-Finch Ranch, to Manila, where, standing on a low bluff as the home of Nels and Mabel Philbrick, it commands much of the same view of the Ballies and Richard's Peak as it formerly did. (Courtesy of Pauline Baker.)

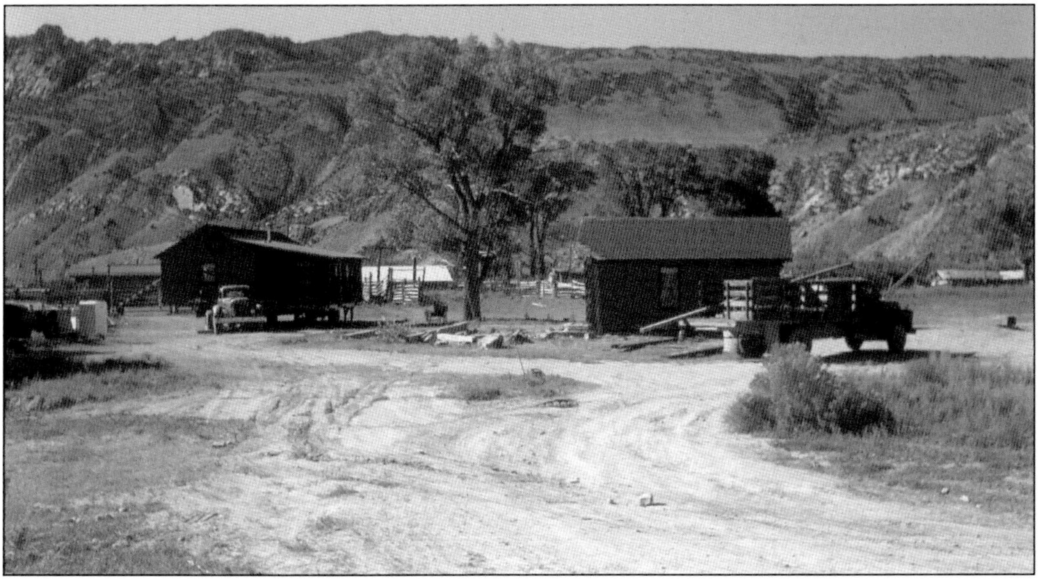

The house on the left is loaded on a truck, ready for the trip to Minnie's Gap, where it will be set on a permanent foundation for Paul and Eleen Williams and their family. (Courtesy of Pauline Baker.)

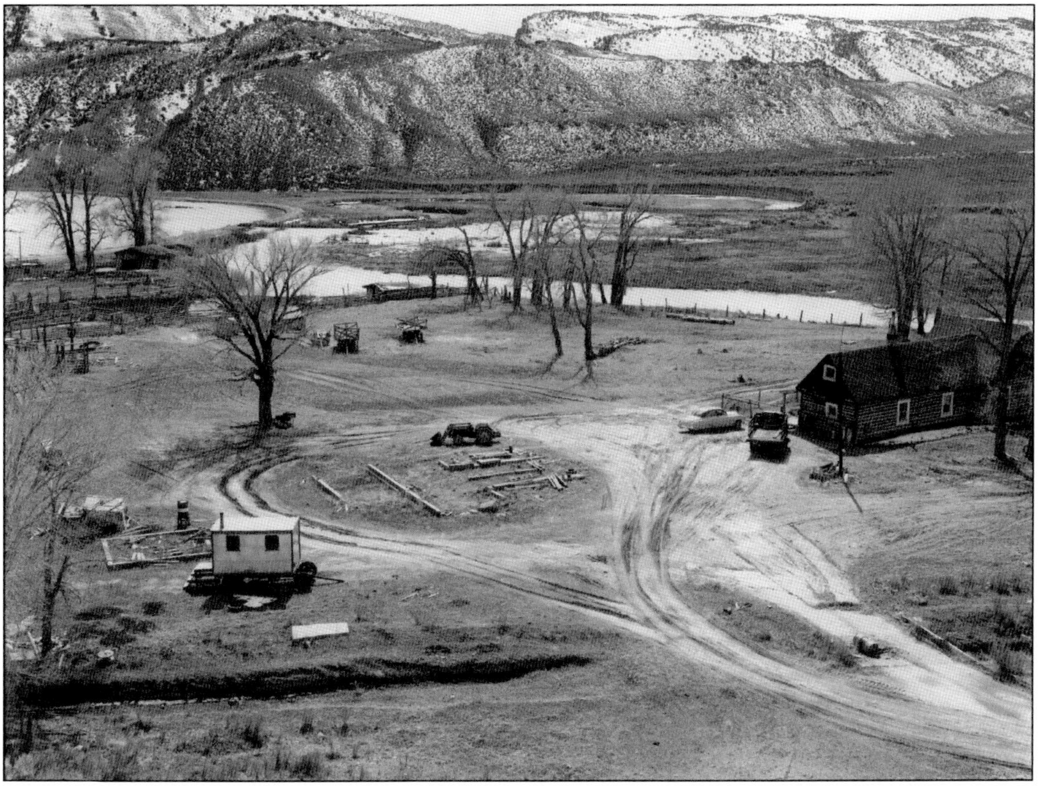

This is the Williams Brothers ranch. The house on the right was moved to Manila. The foundation seen in the center was the site of the Paul Williams home, which was moved to Minnie's Gap. The gravel from the pit in the upper left was used for the dam construction. (Courtesy of BOR.)

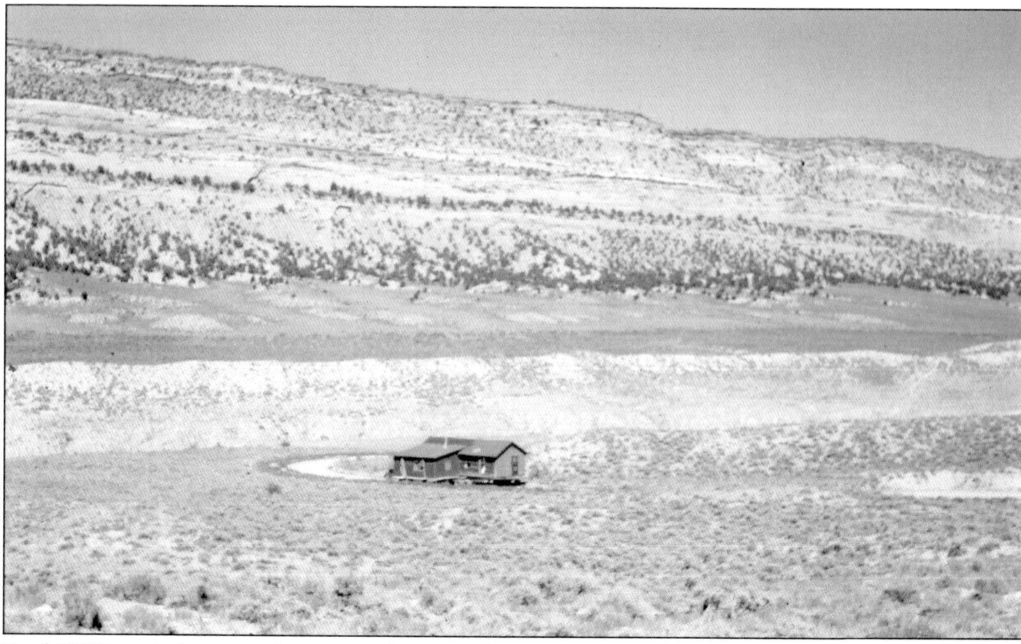

Paul Williams's house is being trucked on the narrow, winding road to Minnie's Gap on the Wyoming side of the Utah-Wyoming state line. (Courtesy of Pauline Baker.)

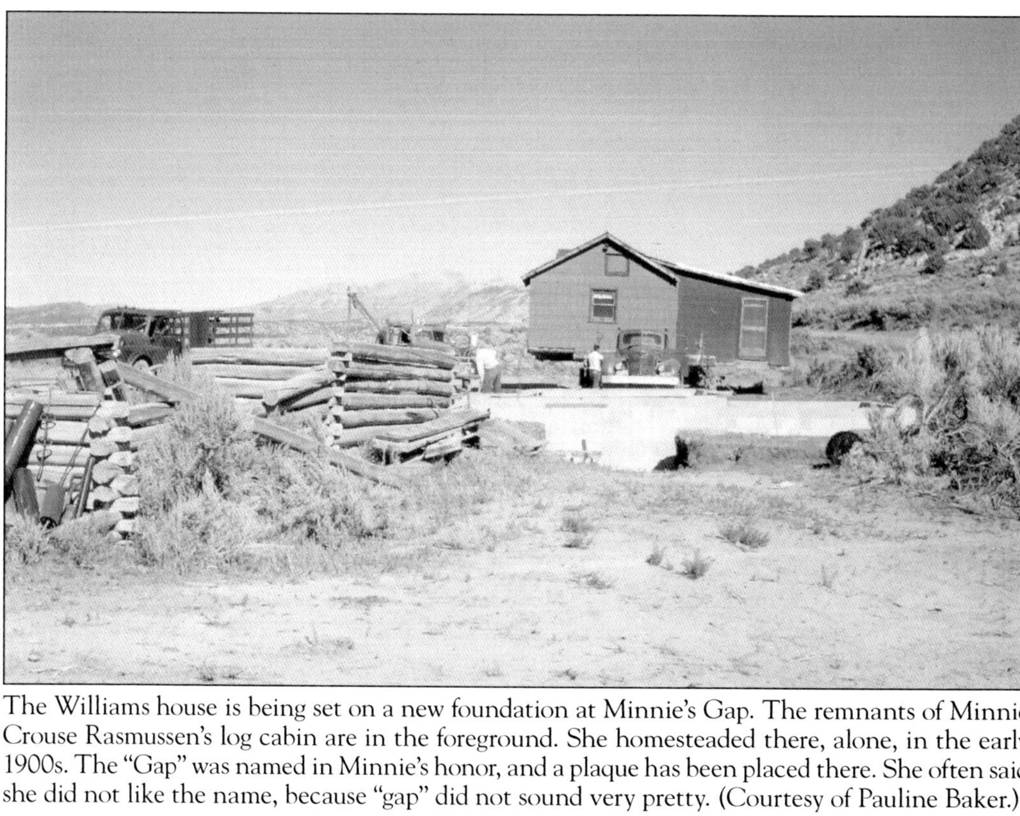

The Williams house is being set on a new foundation at Minnie's Gap. The remnants of Minnie Crouse Rasmussen's log cabin are in the foreground. She homesteaded there, alone, in the early 1900s. The "Gap" was named in Minnie's honor, and a plaque has been placed there. She often said she did not like the name, because "gap" did not sound very pretty. (Courtesy of Pauline Baker.)

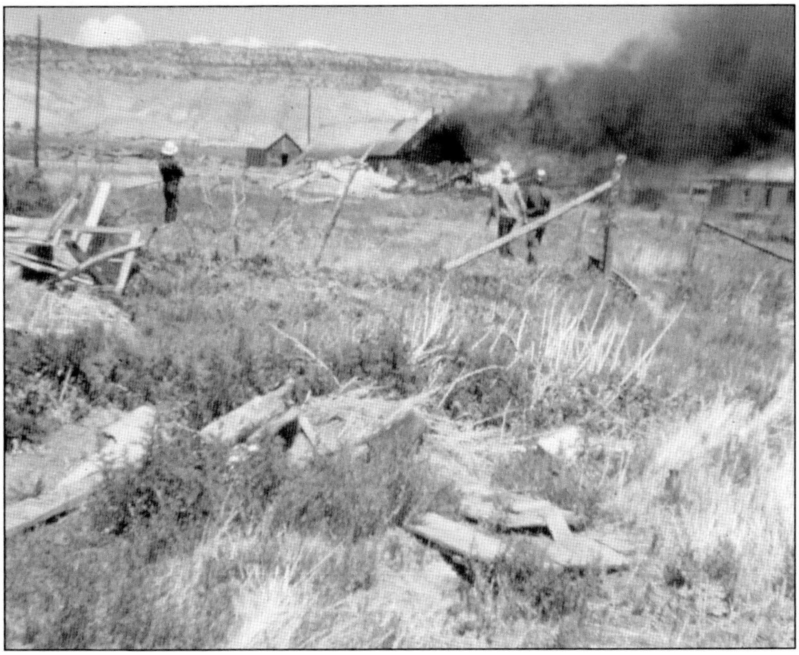

The *Vernal Express* newspaper reported on August 8, 1963: "A combination of water and fire sounded the death knell of tiny Linwood. . . . The water is slowly approaching from the Flaming Gorge reservoir and the fire was ignited last week as part of a clearing operation for the land which will be at the bottom of the reservoir." (Courtesy of Pauline Baker.)

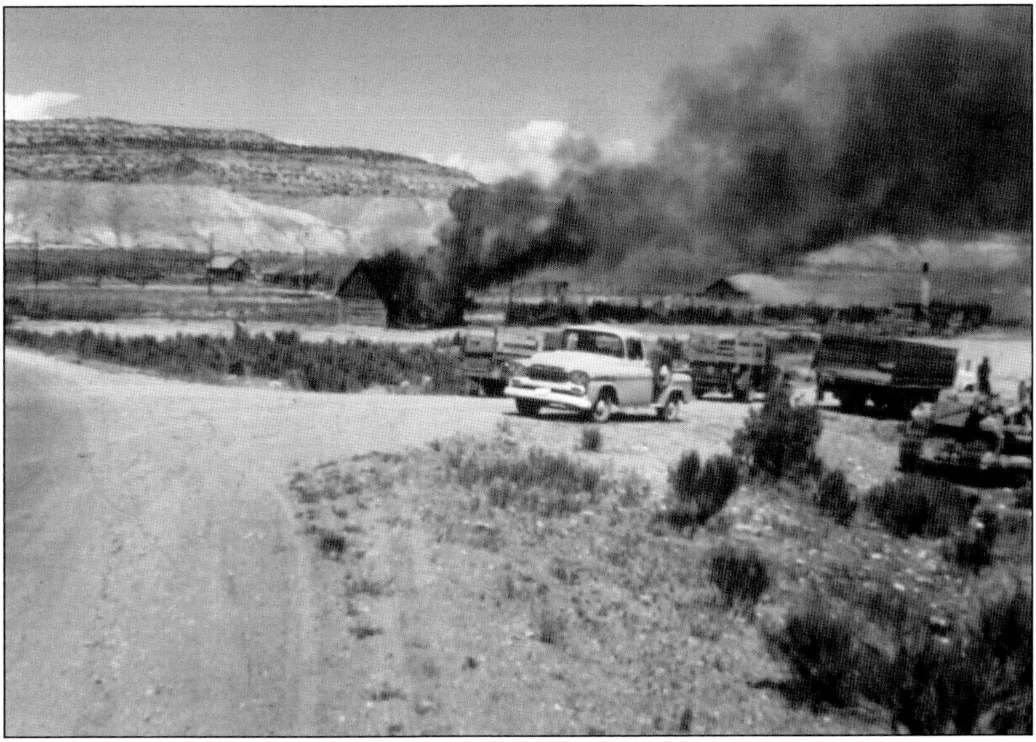

Carma Potter McDowell, who witnessed the clearing of Linwood, recalled: "July 1963, Rich and I climbed up the hillside of Linwood and sadly watched the big cats push in the buildings . . . and then torch each one, and then burning the only home place I had ever known. But life moves on . . . so I tucked each precious memory deep into my heart to hold on to and never let go." (Courtesy of Pauline Baker.)

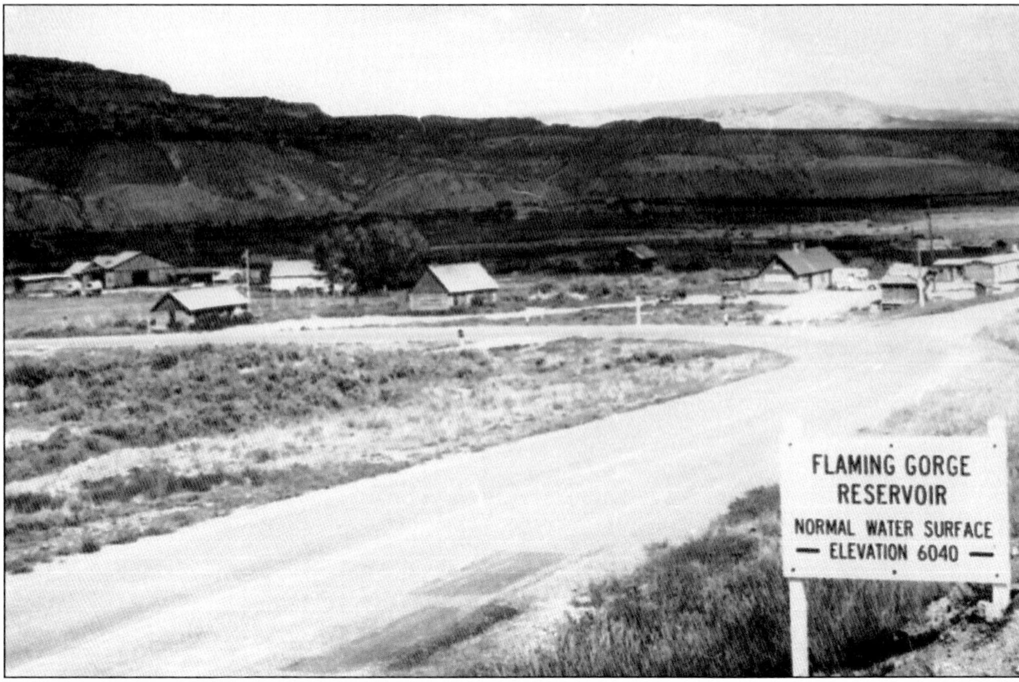

Linwood will be buried deep under water. The dam gates are scheduled to close in November. The store has ceased to operate and will soon be torn down. Other buildings, such as the post office, warehouses, and residences, will disappear. (Courtesy of Pauline Baker.)

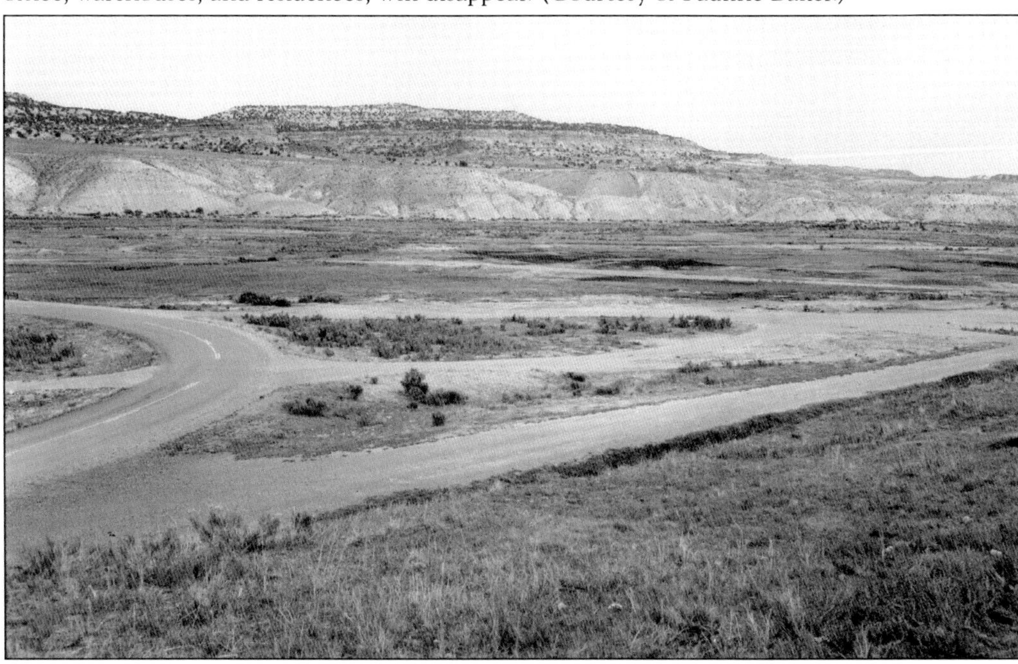

Linwood has vanished so completely that it is hard even for old-timers to remember exactly where it was, but the old town still lives on in spirit. The dead were not forgotten. A good many people were buried in private cemeteries on the ranches where they lived. All of the old graves that could be located were moved up to the cemetery at Manila. (Courtesy of BOR.)

Four

A Bridge Suspended

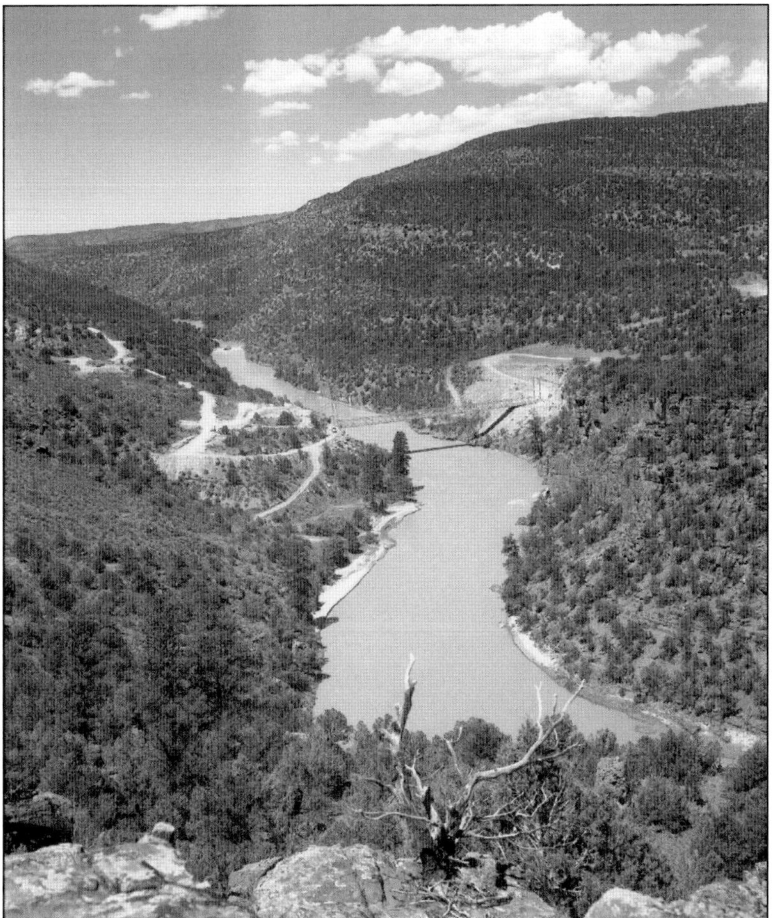

The suspension bridge was a vital link for the construction of the Flaming Gorge Dam, connecting the new community of Dutch John, Utah, with Vernal, Utah. The bridge cut the travel time between the two towns in half. Equipment prepared the roads for the passage of cars and trucks used in the construction of the dam. (Courtesy of BOR.)

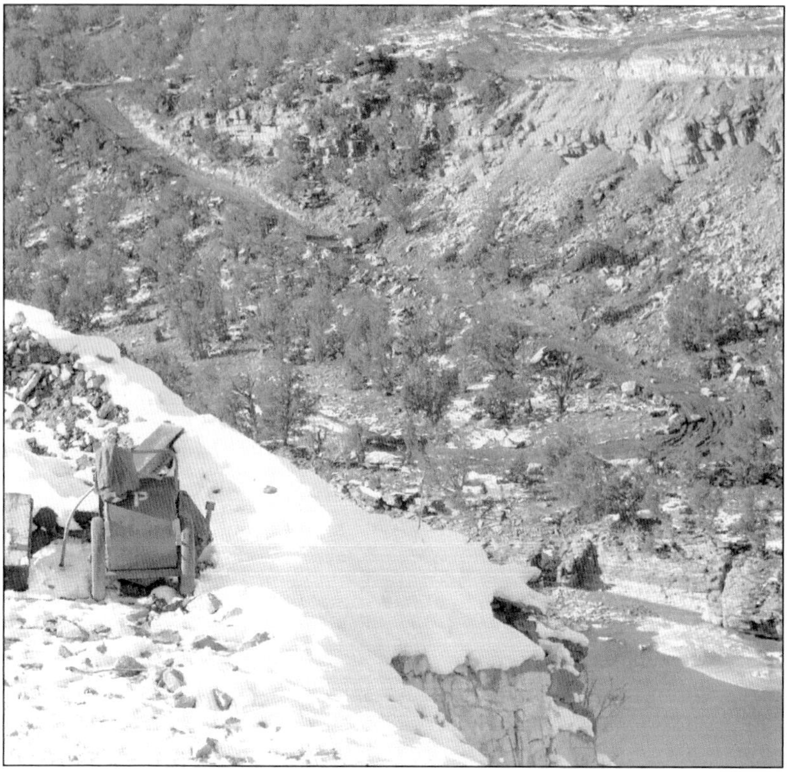

The bridge access road at Flaming Gorge has been graded and graveled. Here, equipment sits on the ledge of the Green River in preparation for roadwork and the bridge to be completed. The temporary road leads to the suspension bridge to be used during construction of the dam and will be the connecting link between Dutch John and Vernal, Utah. (Courtesy of UCRHC.)

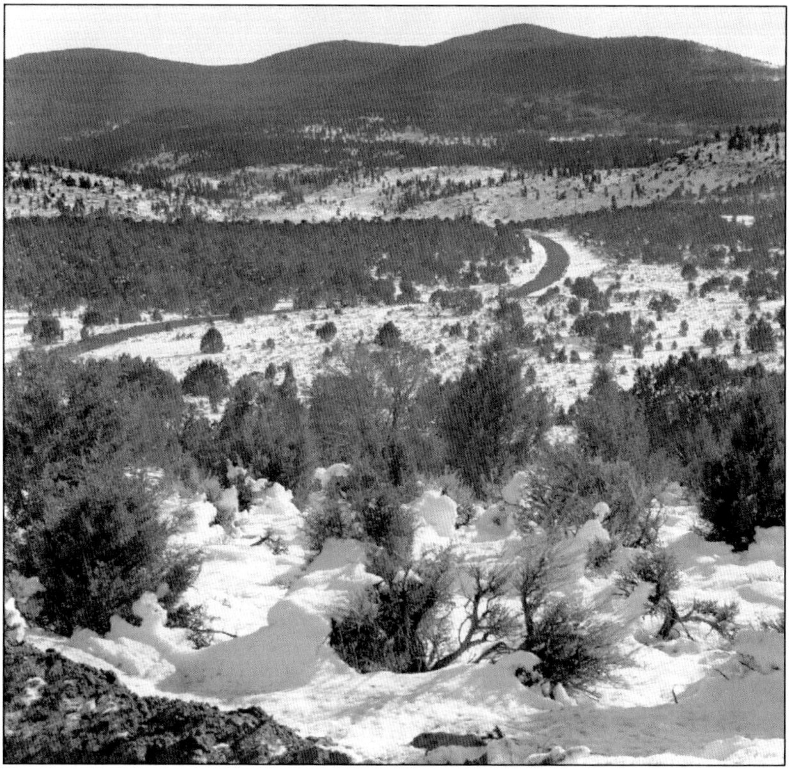

Snowfall at the bridge site does not hinder work on the bridge, which will be located a mile and a half above the Flaming Gorge Dam site. This bridge will carry traffic during the dam's construction, reducing by half the distance for the residents of Dutch John to travel to Vernal, from 98 miles to 47 miles. (Courtesy of UCRHC.)

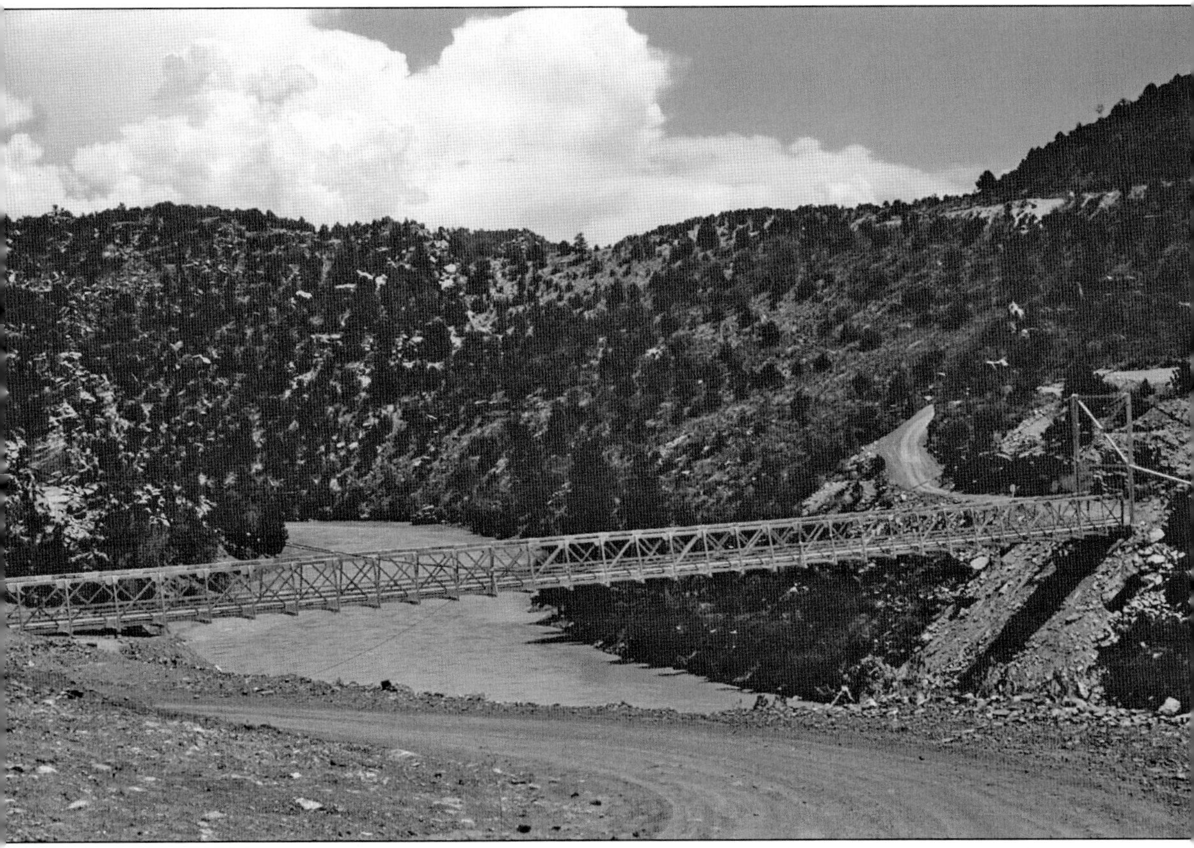

Prior to the Cart Creek Bridge, the connecting link between Dutch John, Utah, and Vernal, Utah, was the suspension bridge. In August 1958, the Utah State Road Commission gave permission for "full speed ahead" for construction of the access road and the suspension bridge. This new bridge and road enabled the residents of the new community of Dutch John to travel to Vernal, Utah, via a shorter route. The Vernal community hoped for a boost to its economy. The bridge, estimated to cost $60,000, will be a 421-foot cable suspension bridge allowing one car at a time to pass. The bridge will be removed to allow the reservoir to fill and to avoid the presence of any remaining debris and material. (Courtesy of Daggett County.)

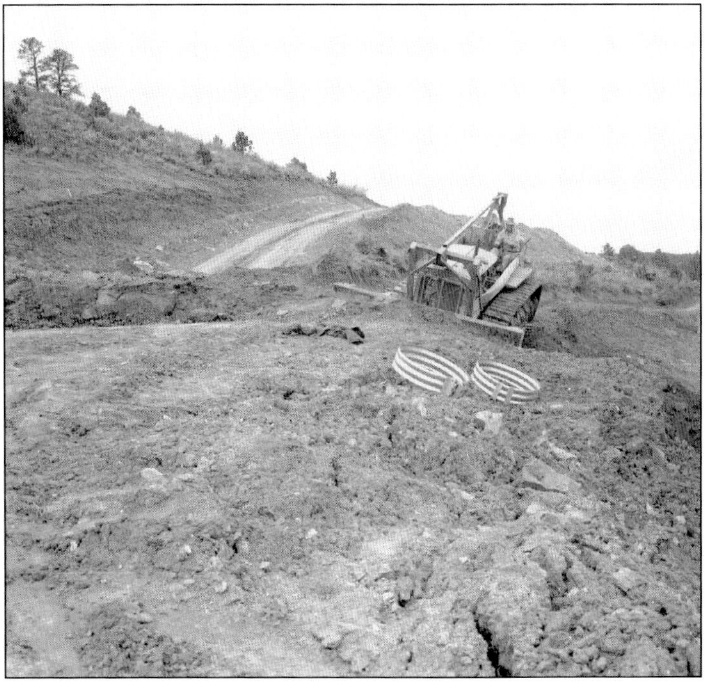

An access road will be constructed down the mountainside to the site of the suspension bridge. This road and the bridge will facilitate travel between Vernal and Dutch John, Utah. Footings for the bridge supports will be built on each side of the Green River. (Courtesy of BOR.)

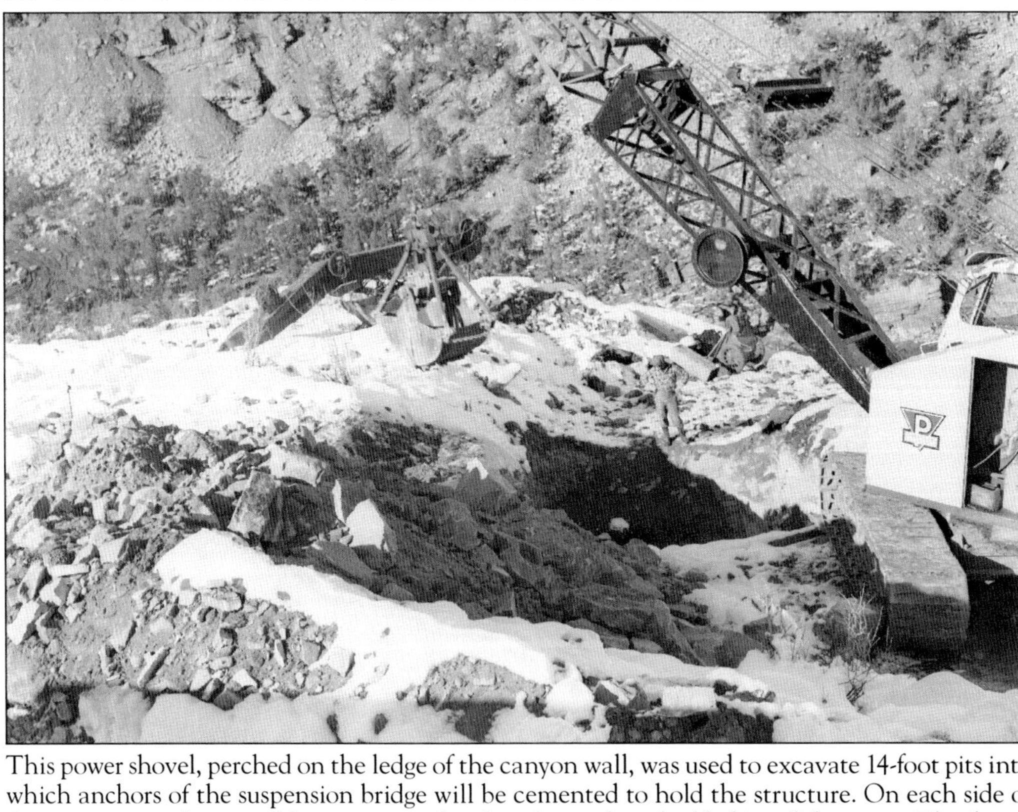

This power shovel, perched on the ledge of the canyon wall, was used to excavate 14-foot pits into which anchors of the suspension bridge will be cemented to hold the structure. On each side of the river, two such pits will be constructed and anchor rods embedded in solid cement. The shovel cleans out pits as material is loosened by blasting and picking. (Courtesy of UCRHC.)

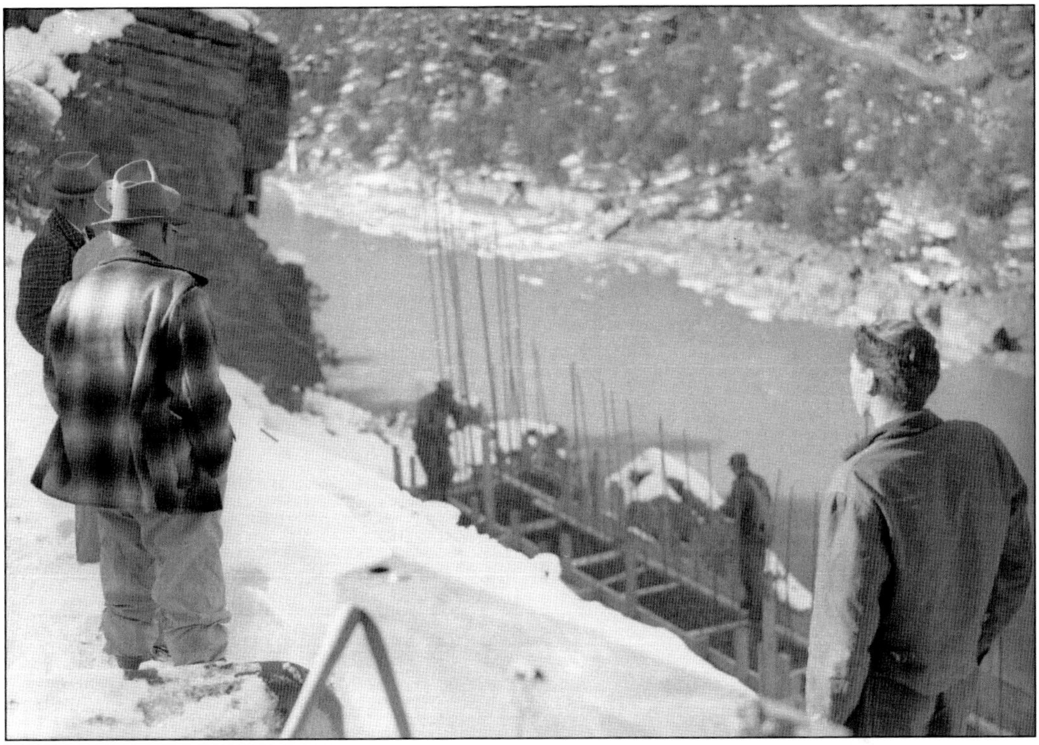

Road chiefs inspect progress on the suspension bridge and the new road. In this view across the Green River from the Greendale side, crews can be seen preparing footings for the bridge tower and anchors. The officials in the foreground worked for the Utah State Road Commission. They are, from left to right, bridge inspector Ed Carman, resident engineer Ross Slye (partially obscured), bridge superintendent E.R. Bethel, and Barry Gale of Vernal. (Courtesy of UCRHC.)

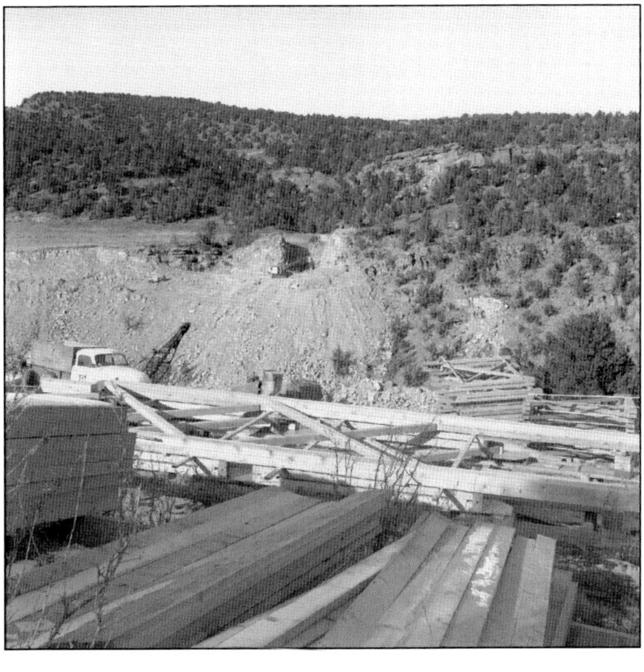

Power shovels and other equipment begin work on the suspension bridge that will be constructed between Angel Falls and the Flaming Gorge Dam site. Holes will be dug for the bridge supports and the grading and graveling of the road finished soon. Cables will cross the river and be the support system for the bridge. Decking and side railings are being gathered for the bridge's construction. (Courtesy of UCRHC.)

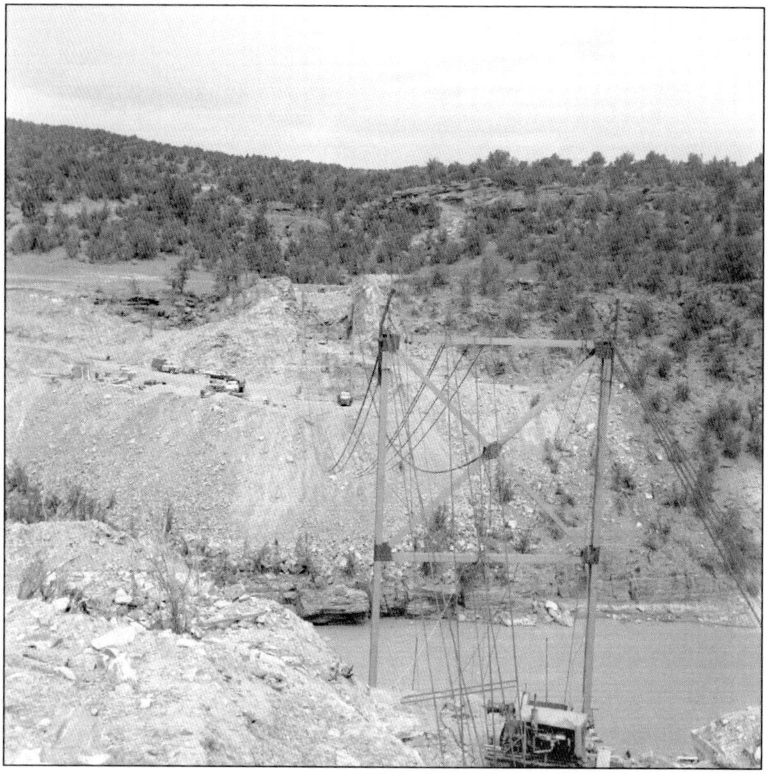

Construction of the suspension bridge required the use of large cables strung across the Green River. The bridge was constructed near Angle Falls, just a mile and a half above the Flaming Gorge Dam site. The bridge shortened by half, and made easier, travel between Dutch John and Vernal. (Courtesy of UCRHC.)

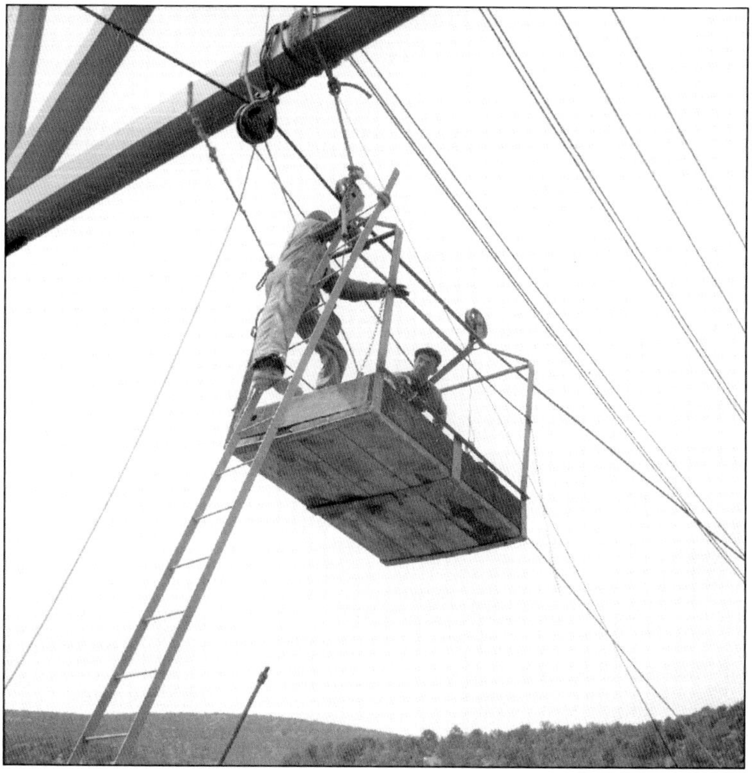

Come aboard! Here, two workers prepare to travel across the Green River. This trolley platform gives the traveler an interesting and somewhat uneasy view of the river and the surrounding country. Its main use was for transporting men and materials across the river during construction of the suspension bridge. (Courtesy of UCRHC.)

Ross Slye (right), the resident engineer for the Utah State Road Commission, speaks with an unidentified man. The suspension bridge is ready for the supports to be added to the sides. Residents of Daggett and Uintah Counties are anxious for this bridge to be finished, as it will shorten the trip between the counties. The bridge will be removed when the Flaming Gorge Dam and the Cart Creek Bridge are completed. (Courtesy of UCRHC.)

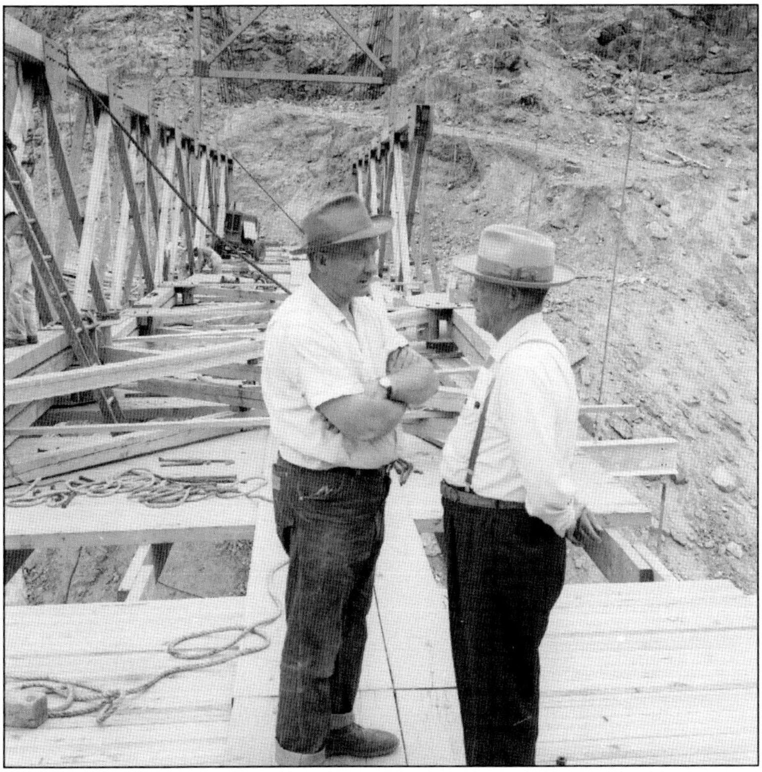

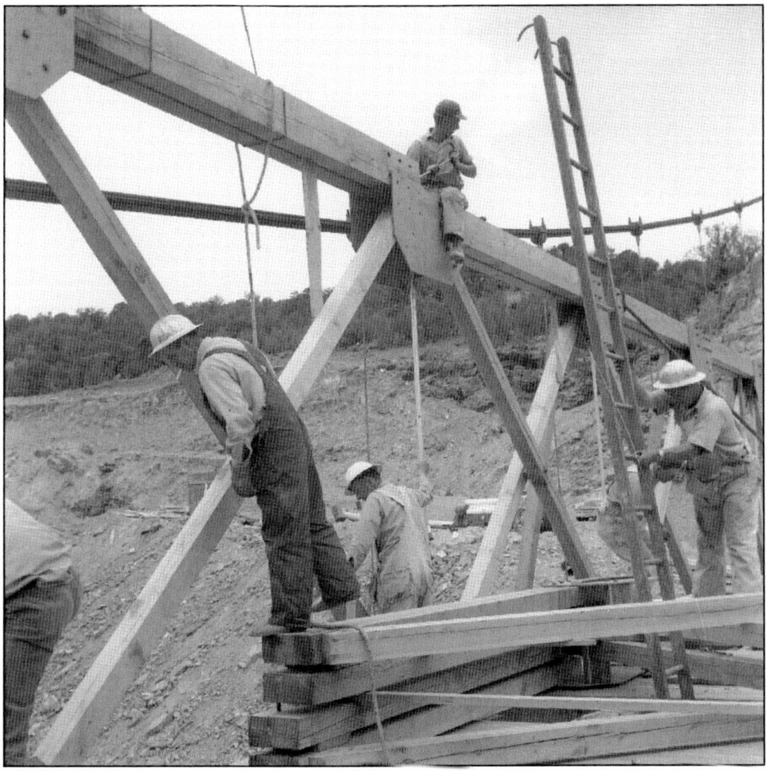

It's a long way down! The bridge crew speeds up work in order to complete the structure by Friday, May 22, 1959. Railing and supports are being installed as the decking is completed. The cable supporting the bridge was assembled in seven sections for extra strength. (Courtesy of UCRHC.)

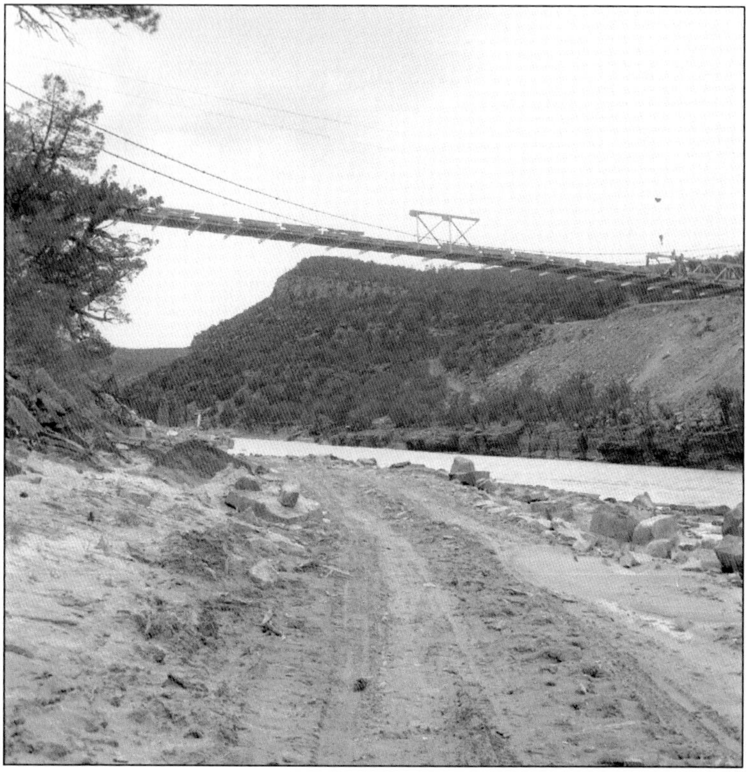

Over the river! Sides of the bridge are going into place rapidly. Looking up from the riverbed, the suspension bridge over the Green River is nearing completion. The bridge was finished in May 1959, and Vernal welcomed the residents of Dutch John with a "Hi Neighbor Day." This bridge was a great asset to the Uinta Basin for residents of the two communities. (Courtesy of UCRHC.)

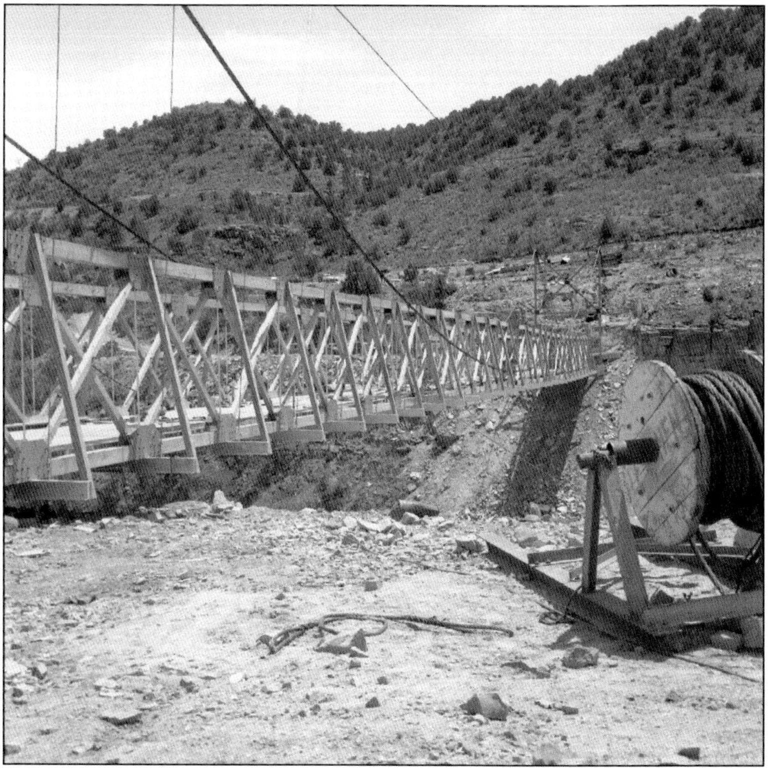

Almost done! Crews work fast to complete the bridge by the end of May. This view is from the east side of the bridge. The decking and sides are in place, and some cleanup work is needed. The bridge was removed when the Flaming Gorge Dam and Cart Creek Bridge were completed. (Courtesy of UCRHC.)

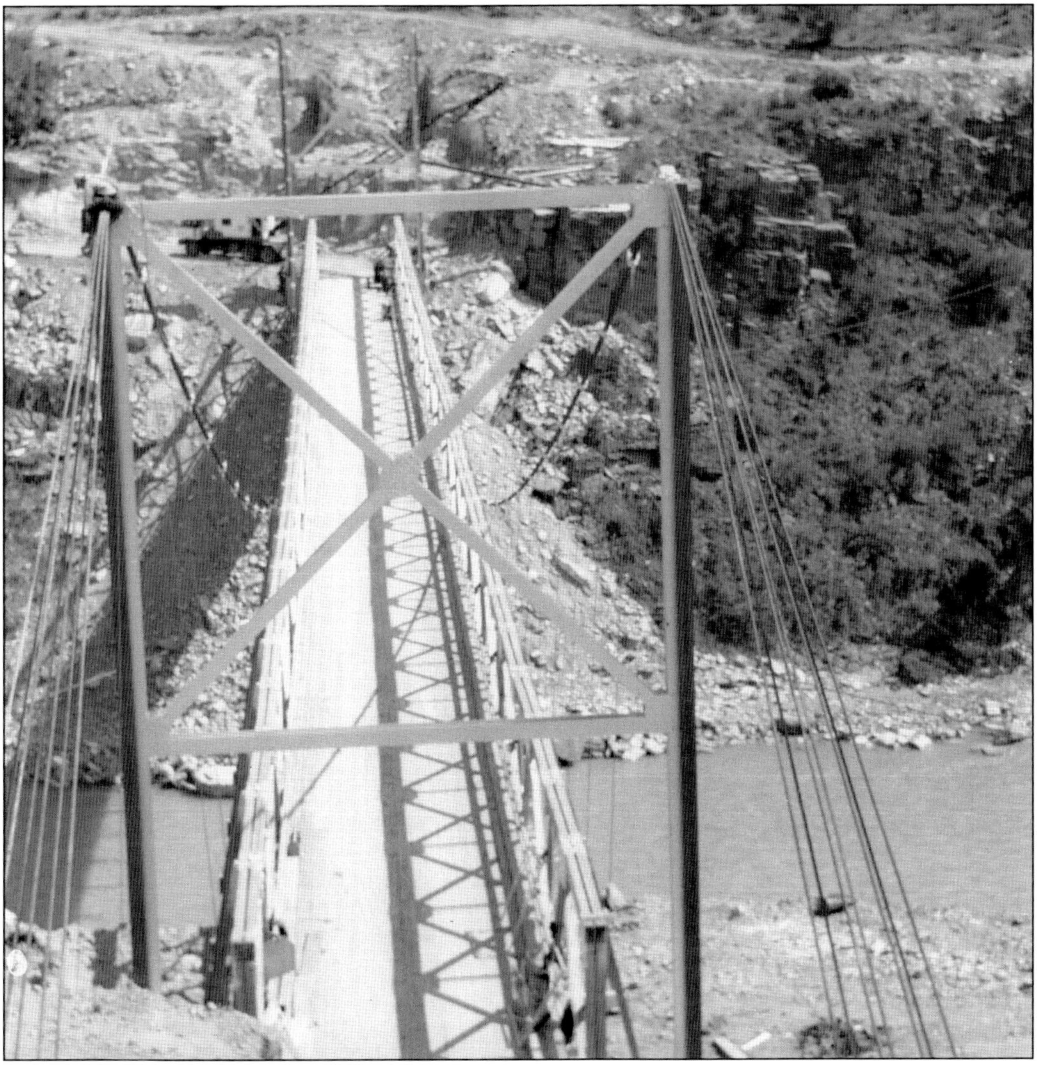

It is done! It is finished! The bridge crew speeds work to complete the structure by Friday, May 22, 1959. The new bridge is sturdy and sound, and inspectors have okayed it for travel and will open it to the public. The cable is assembled in seven sections for extra strength. The 421-foot-long cable suspension bridge was completed at a cost of $118,653, significantly more than the estimated $60,000. Folks used this bridge for just a short time until the Cart Creek Bridge was completed. (Courtesy of UCRHC.)

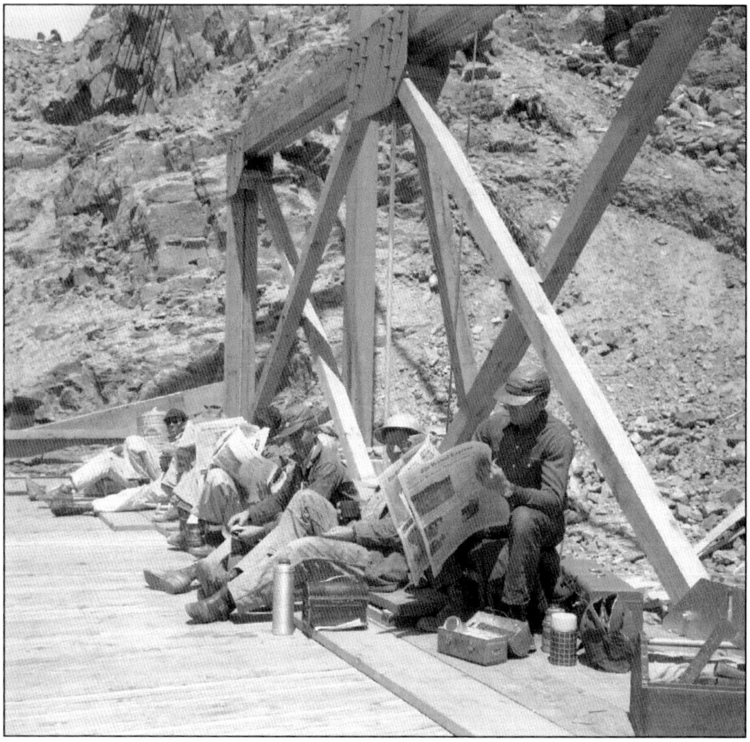

Lunchtime! Workers on the access bridge take a well-deserved lunch break to eat and read the special edition of the *Vernal Express* commemorating the dedication of the bridge and outlining the free events for Daggett County residents. The dedication was on June 6, 1959, and the program lasted 30 minutes. (Courtesy of UCRHC.)

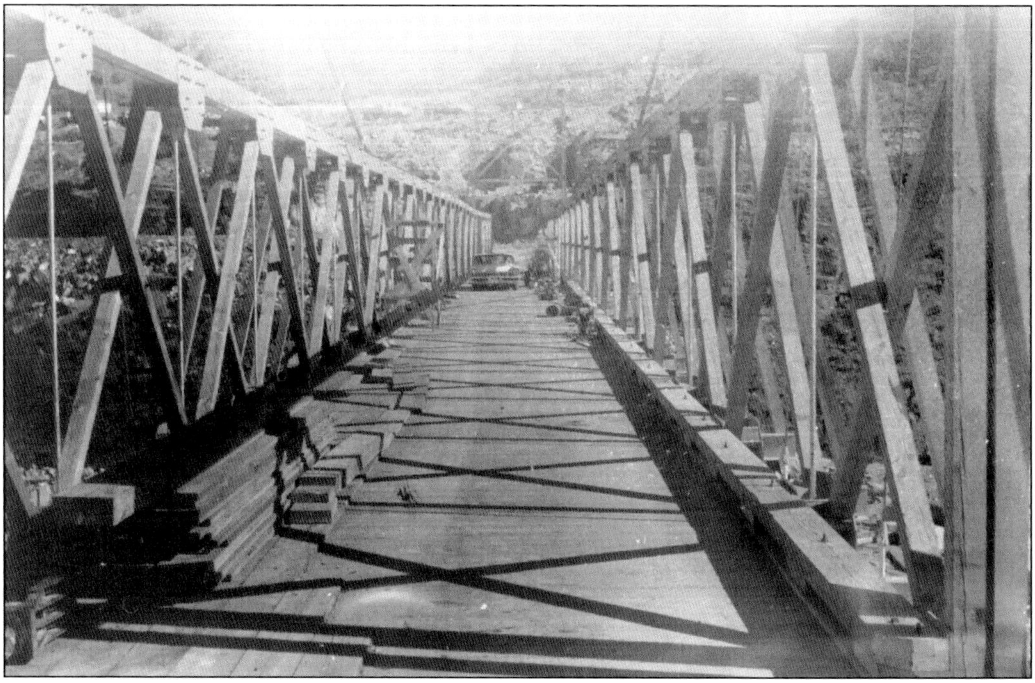

One of the first cars to cross the bridge between Angel Falls and the construction site of the Flaming Gorge Dam does so slowly, as it has to maneuver around construction material left on the bridge. This span served residents of Dutch John and Vernal, allowing the communities to access one another in a shorter distance of time. (Courtesy of UCRHC.)

Five

Golden Gate of the Uintas

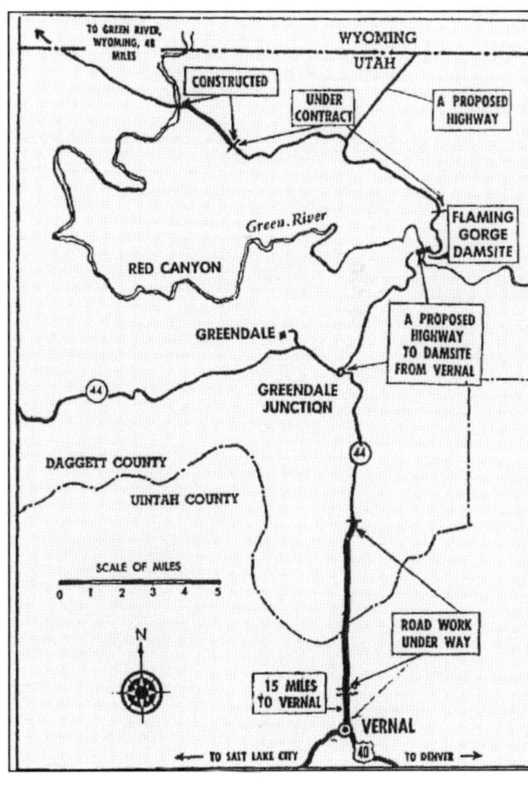

The new roads built during the construction of the Flaming Gorge Dam provided vital links for Dutch John and Vernal. But, a big problem faced the Utah Road Commission during the construction, with the need for a permanent road connecting the dam with Wyoming. A new highway was proposed between Greendale Junction and the Flaming Gorge Dam site that would then run across the dam and into the new community of Dutch John, Utah. (Courtesy of UCRHC.)

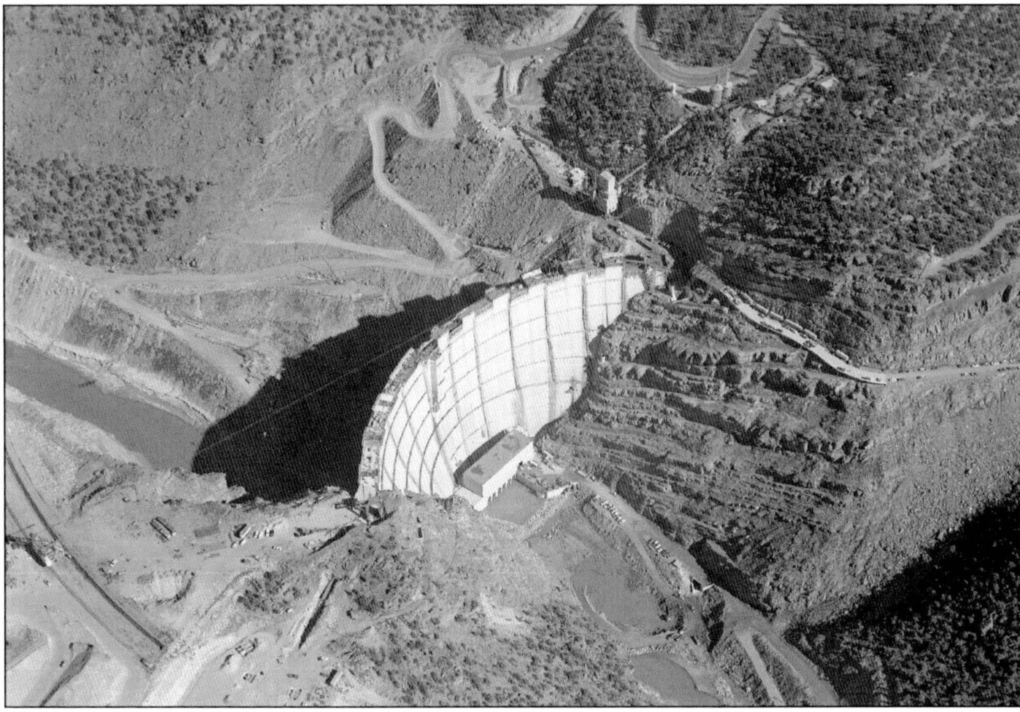

Construction progress is evident in this aerial photograph of the Flaming Gorge Dam and power plant. The highway will come from the Cart Creek Bridge, down and across the dam, and around the cement road being built against the mountains. This is part of the highway that will take traffic from Greendale Junction to Dutch John, Utah. (Courtesy of BOR.)

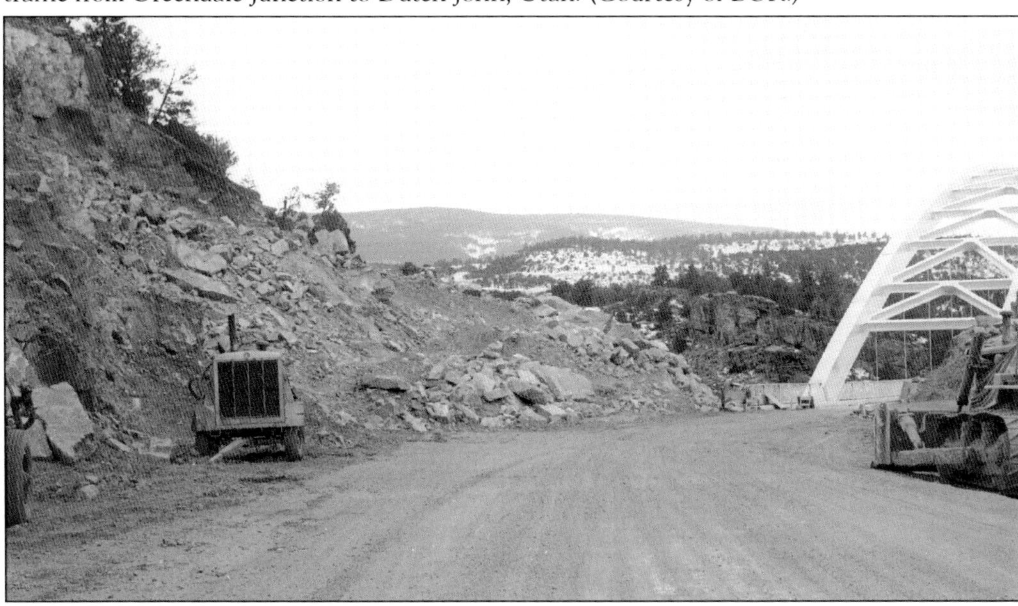

Heavy rains have caused the soil and rocks to slowly slide toward the new Cart Creek Bridge. The contractor is working to clear the boulders and debris from the road and to keep the bridge from being damaged. The finished highway will run across the bridge, over the dam, and to Dutch John, Utah. (Courtesy of UCRHC.)

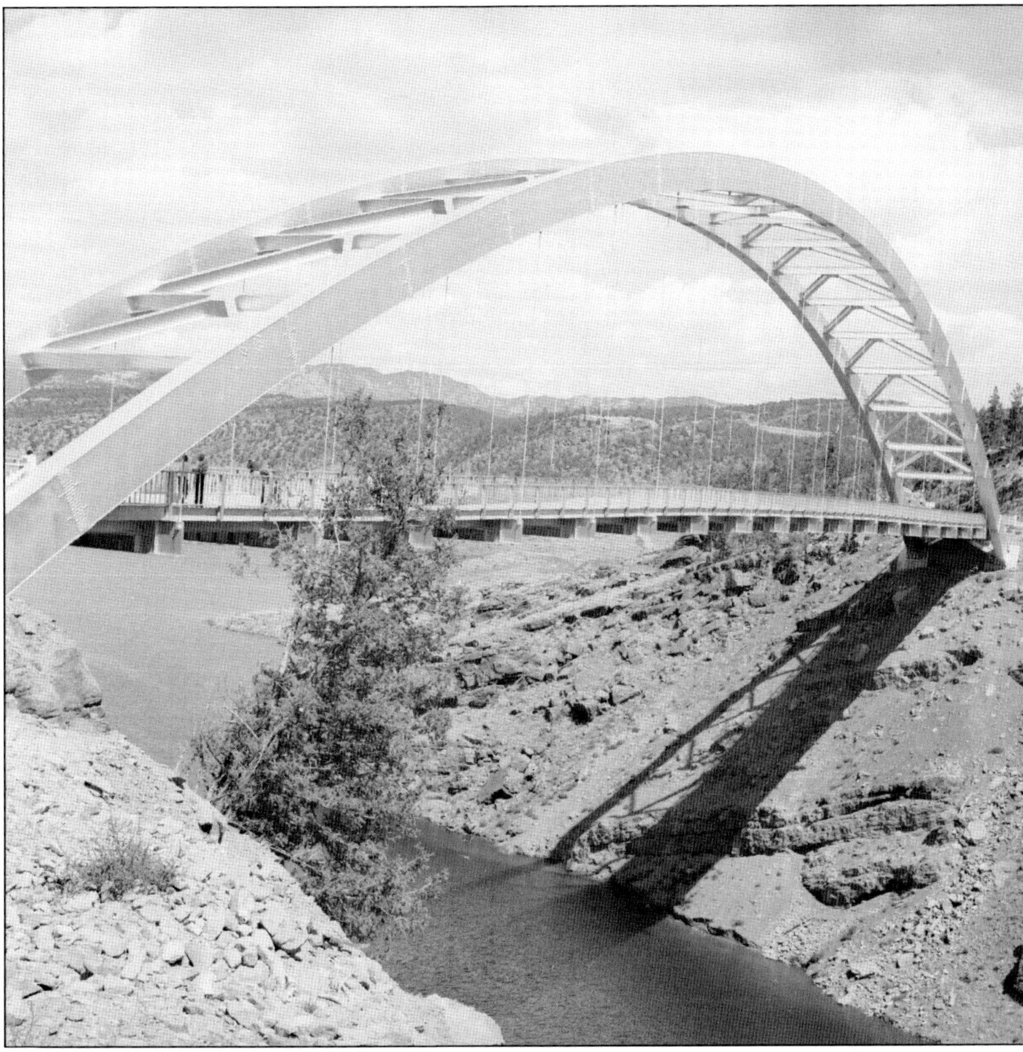

The contract for the Cart Creek Bridge was awarded to US Steel, American Bridge Division, of Denver, Colorado, at a low bid of $768,782. The bridge will be the longest single-span in the state of Utah at 550 feet. State officials gathered in December 1962 for the dedication. Built for a total cost of $810,000, the bridge was a link to Flaming Gorge for construction workers of the dam. Cart Creek Bridge won an award of merit from the American Institute of Steel Construction for its beauty, and its graceful arch served as a model for two other Utah bridges. (Courtesy of UCRHC.)

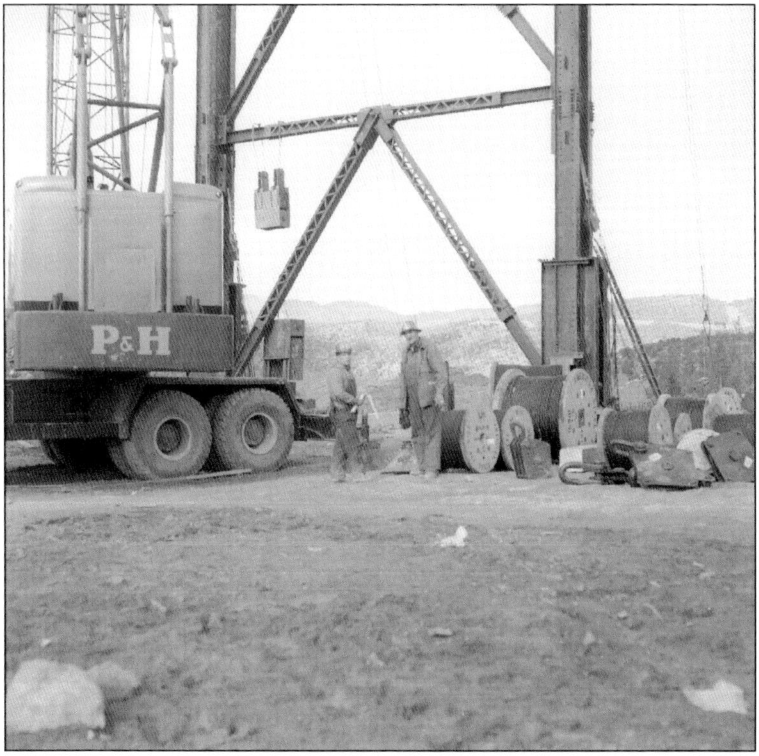

Cloyd McCarrell (left) and G.L. McLure stand beneath the tower used to construct the arch of steel at Cart Creek Gorge. This structure will be the support device for the arches placed over the canyon. It will be removed after the bridge is completed. (Courtesy of UCRHC.)

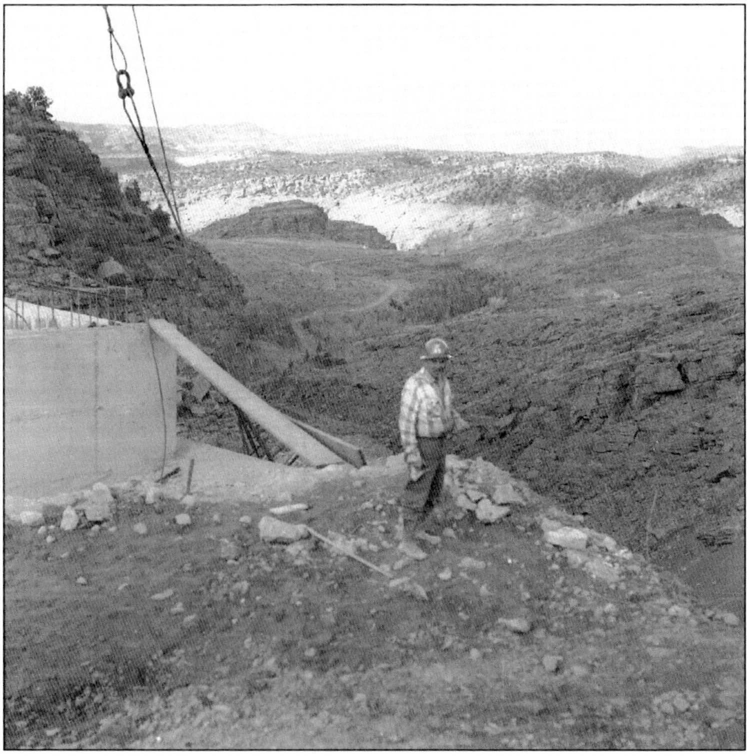

Superintendent A.E. Wotring of American Bridge, a division of US Steel, stands at Cart Creek Gorge, where the Cart Creek Bridge is under construction on Flaming Gorge Highway. To the left are the cement foundations that will be used for the arch supports. (Courtesy of UCRHC.)

This is one of the steel support structures that will be used to construct the arch across the Cart Creek Gorge. The structure is 150 feet high; when the jib is extended, it is 160 feet high. There will be one of these on each side of the gorge, serving as a support for the arches of the final bridge. They will be removed when the bridge is completed. (Courtesy of UCRHC.)

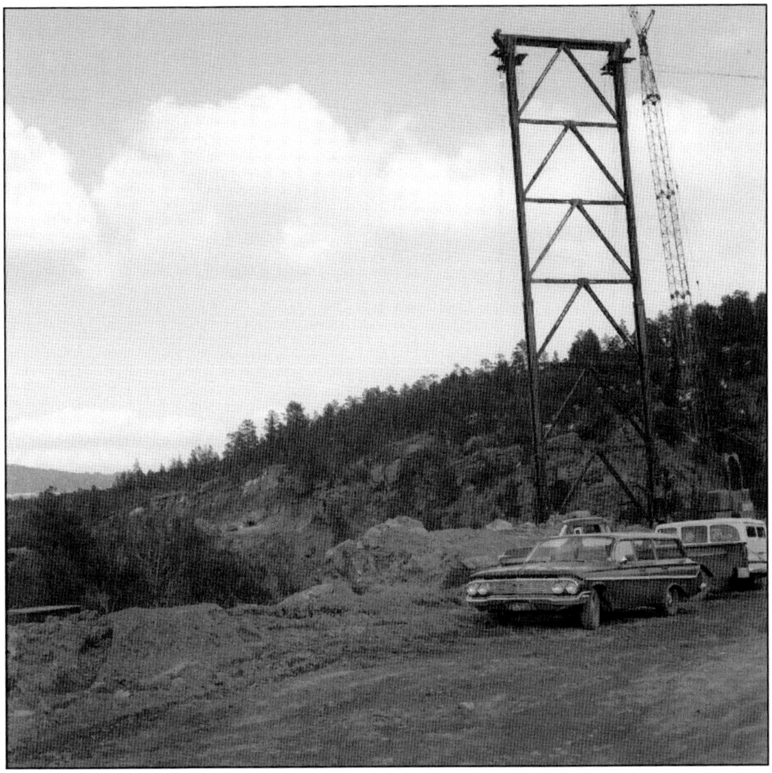

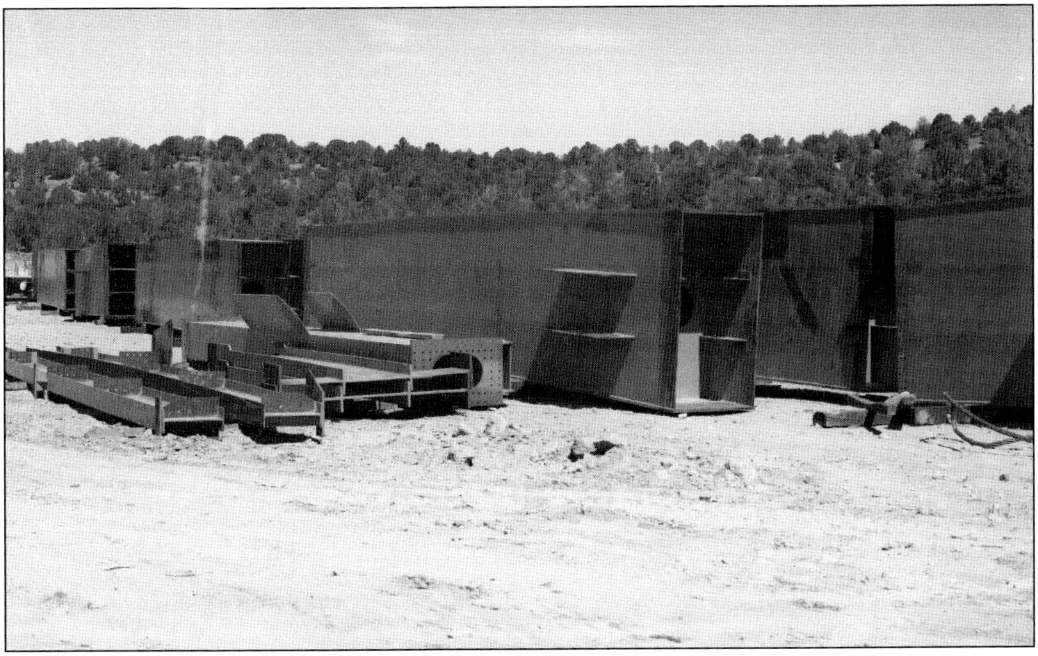

Steel boxes brought to the Cart Creek Bridge location were used to build the arch. The box sections of the welded steel plates, measuring about 42 inches wide and over seven feet deep, were pieced together to form the arch. The steel sections varied in length, from 25 to 40 feet, and weighed between 11 and 22 tons. (Courtesy of UCRHC.)

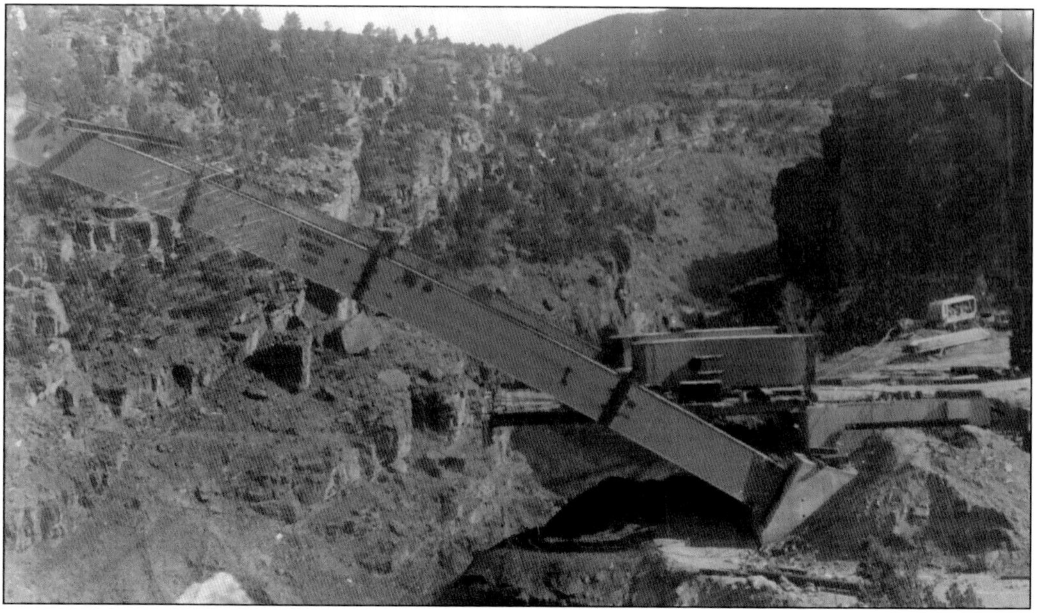

Bridge builders set the first box section of the arch. They will join 20 other units to form the structure to support the bridge. The box sections of the welded arches, measuring about 42 inches wide and over seven feet deep, are being attached to create the arch for the Cart Creek Bridge. The steel sections vary in length from 25 to 40 feet and weigh between 11 and 22 tons. (Courtesy of UCRHC.)

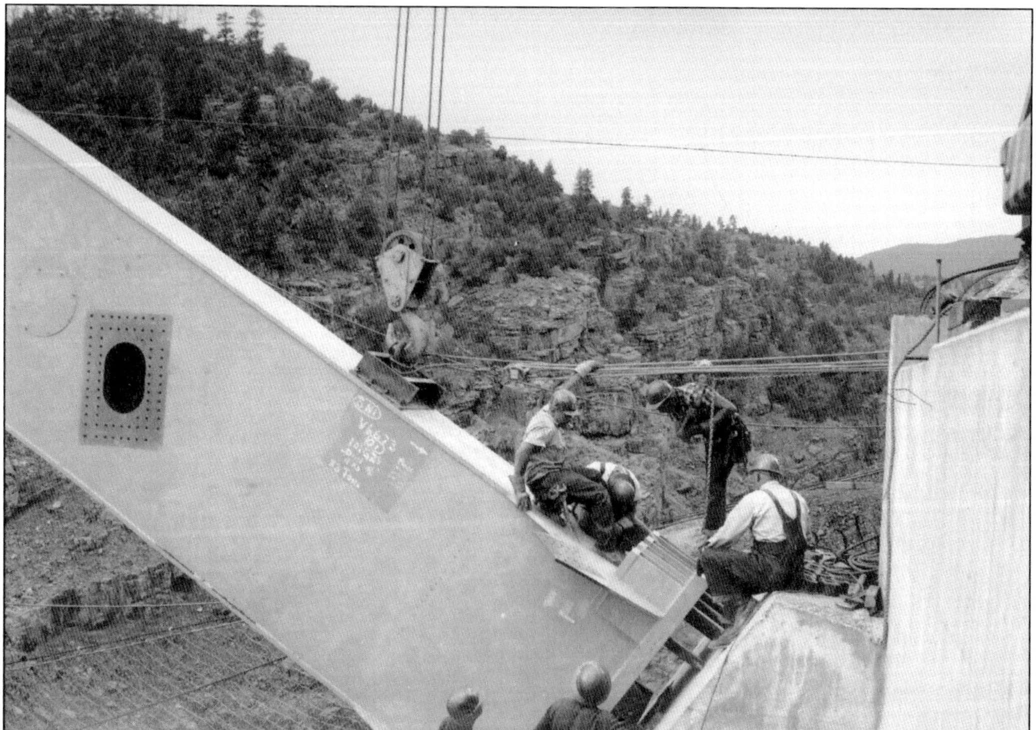

Construction work is proceeding rapidly on the Cart Creek Bridge. Massive steel arches hang over the 200-foot-deep Cart Creek Canyon. Midway between the end abutments of the bridge, the ribs will arch to their highest point, 90 feet above the deck. When completed on November 1, at 550 feet, it will be the longest single-span bridge ever built in Utah. (Courtesy of UCRHC.)

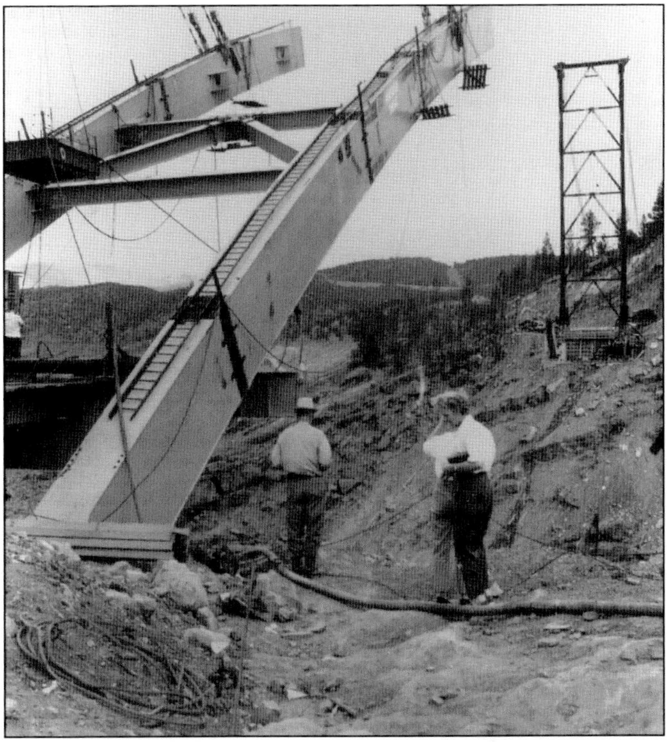

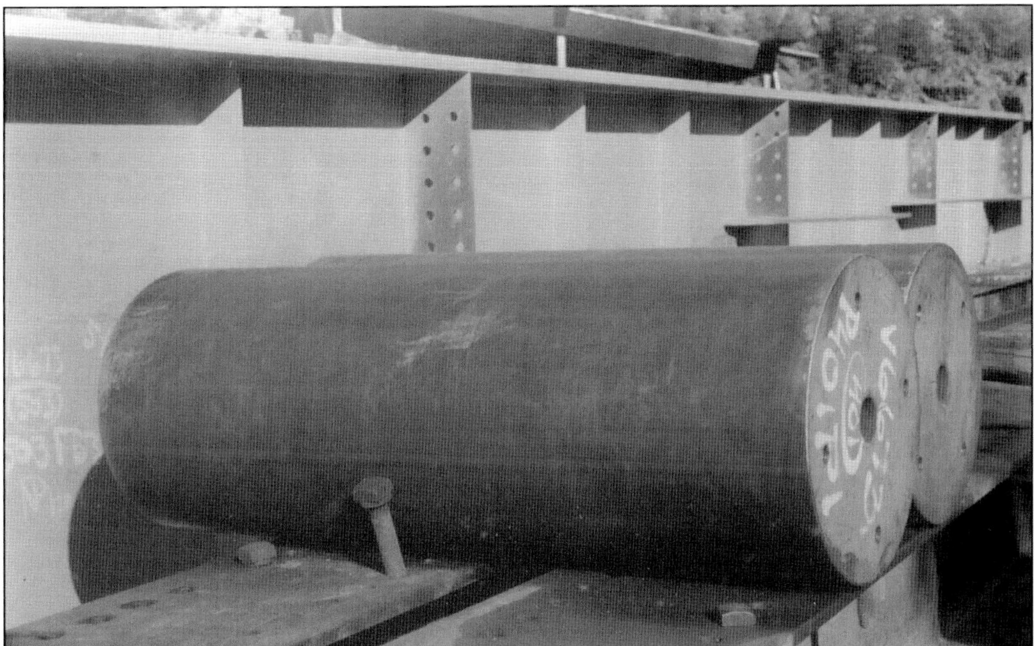

The two massive arched sections of the Cart Creek Bridge will be joined by these large bearings, which resemble bombs. They will be put into place in August 1962, after which work will begin on the suspension system. To be installed are 1 3/4-inch steel hangers to support Utah's longest single-span bridge, a two-lane roadway. (Courtesy of UCRHC.)

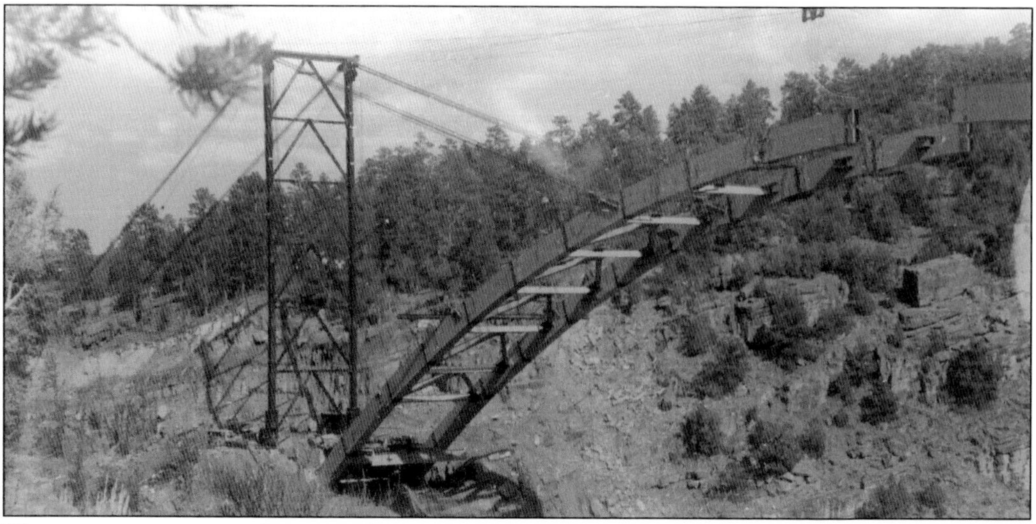

The two enormous arches joined "arms" on August 31, 1962. Crews, using intricate cable arrangements, eased the two 15-ton sections of the bridge together. Following this, work began on the suspension system, installing the deck for the two-lane roadway between Vernal and Flaming Gorge. (Courtesy of UCRHC.)

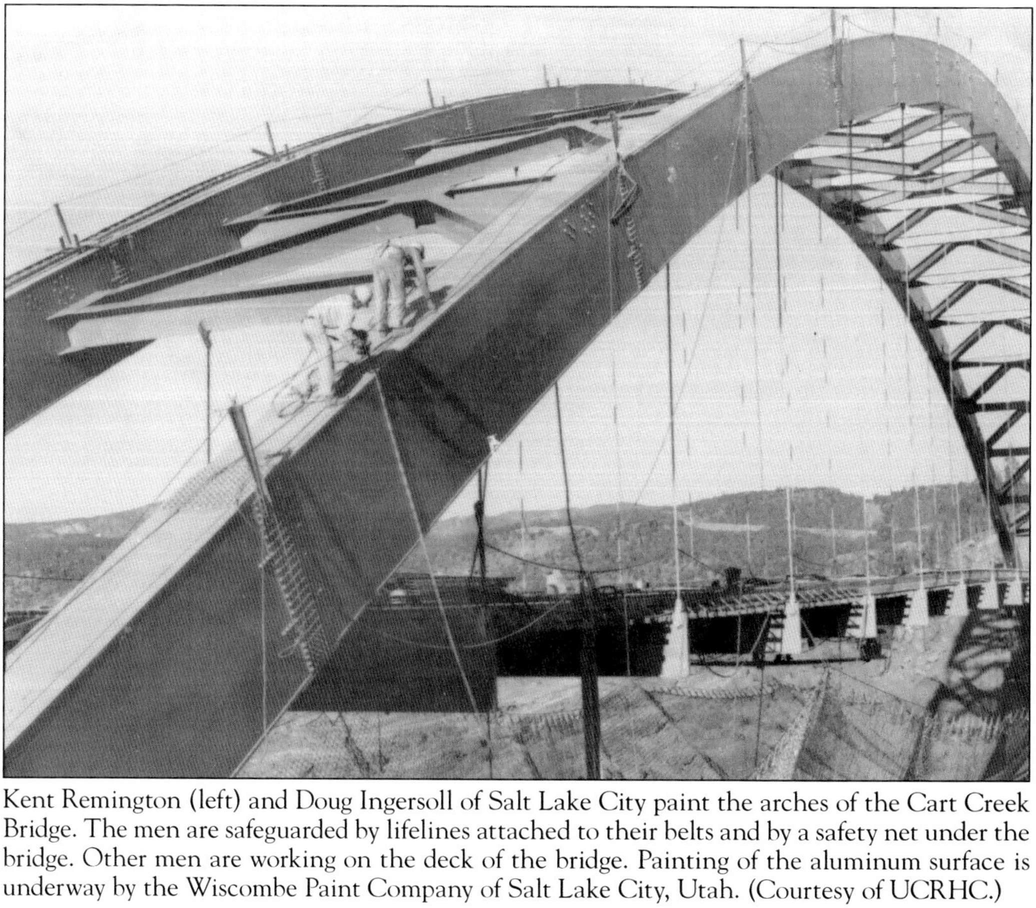

Kent Remington (left) and Doug Ingersoll of Salt Lake City paint the arches of the Cart Creek Bridge. The men are safeguarded by lifelines attached to their belts and by a safety net under the bridge. Other men are working on the deck of the bridge. Painting of the aluminum surface is underway by the Wiscombe Paint Company of Salt Lake City, Utah. (Courtesy of UCRHC.)

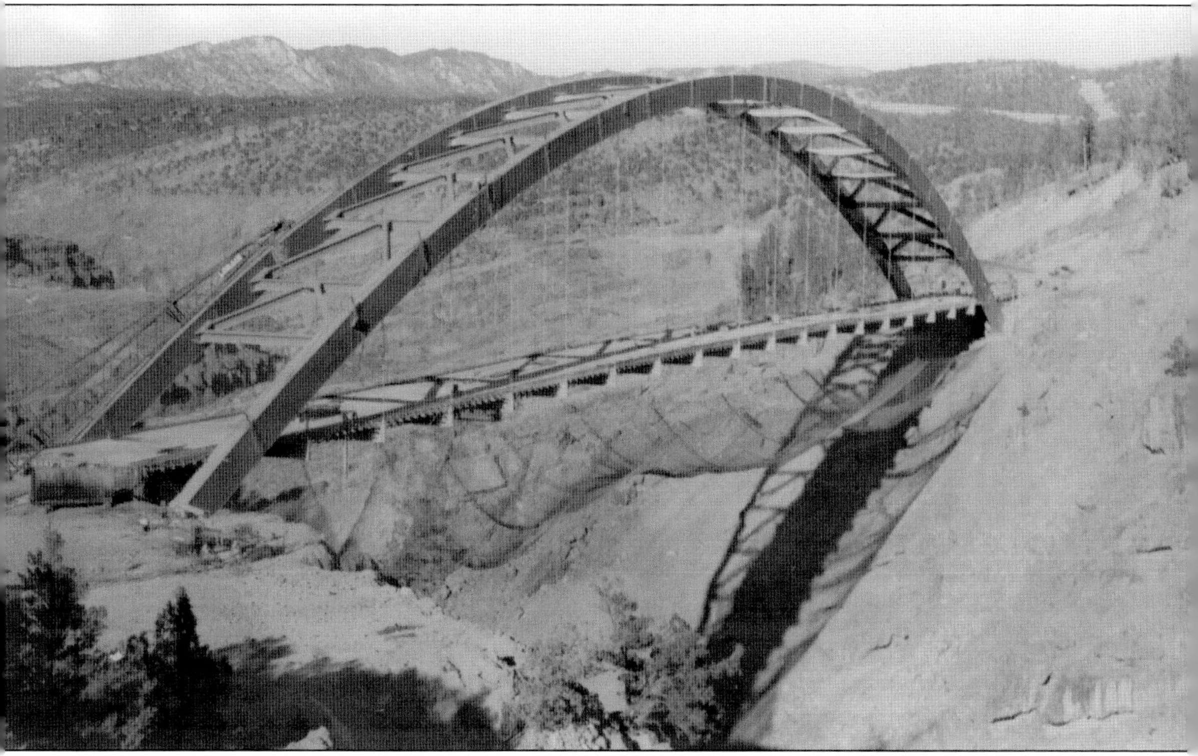

The massive arch extends over the gorge in this photograph taken on October 10, 1962. The early afternoon sun is throwing deep shadows of the steel arch and deck of the Cart Creek Bridge upon the canyon wall. The workers are protected by the safety nets under the bridge. The crew had put in all the steel, decking, and approaches. The American Bridge division of US Steel would later install handrails and do the finishing work. The cost of the structure came to about $800,000. The highway approach to the bridge from either end was finished by the Utah State Road Commission. (Courtesy of UCRHC.)

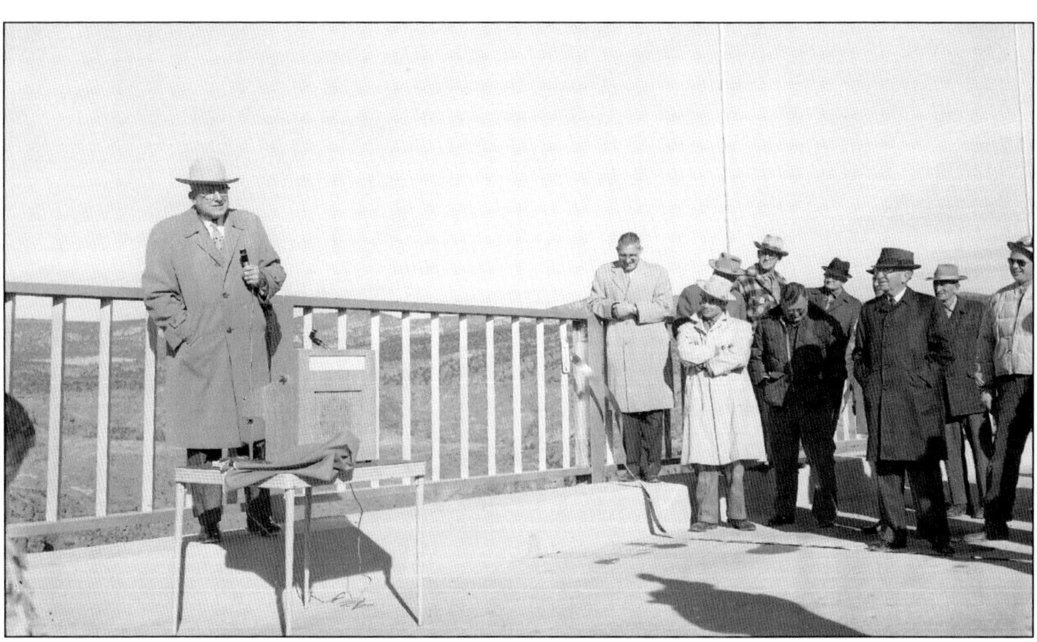

On December 14, 1962, Utah's outstanding Cart Creek Bridge was officially dedicated. George D. Clyde, governor of Utah, spoke to nearly 300 people gathered in the remote northeastern corner of the state. Among the prominent men in the photograph above are Royal Henderson (leaning on rail), Fran Feltch (light coat and hat), Grant Southam (head bowed), and Bry Stringham (dark coat and hat). The unusual bridge, which claimed a number of "firsts," was designed by D.L. Sargent, chief structural engineer of the Utah State Department of Highways. A.E. Wotring served as American Bridge construction superintendent. Below, participating in the dedication is the Manila High School band, which provided enthusiastic music as its members marched briskly across the newly completed span. (Courtesy of UCRHC.)

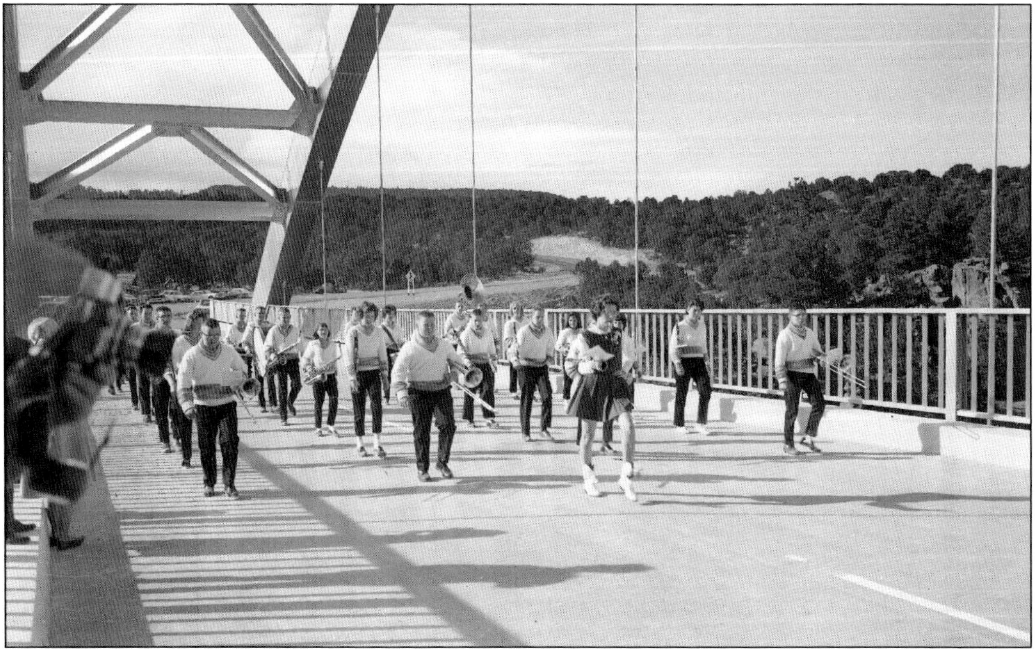

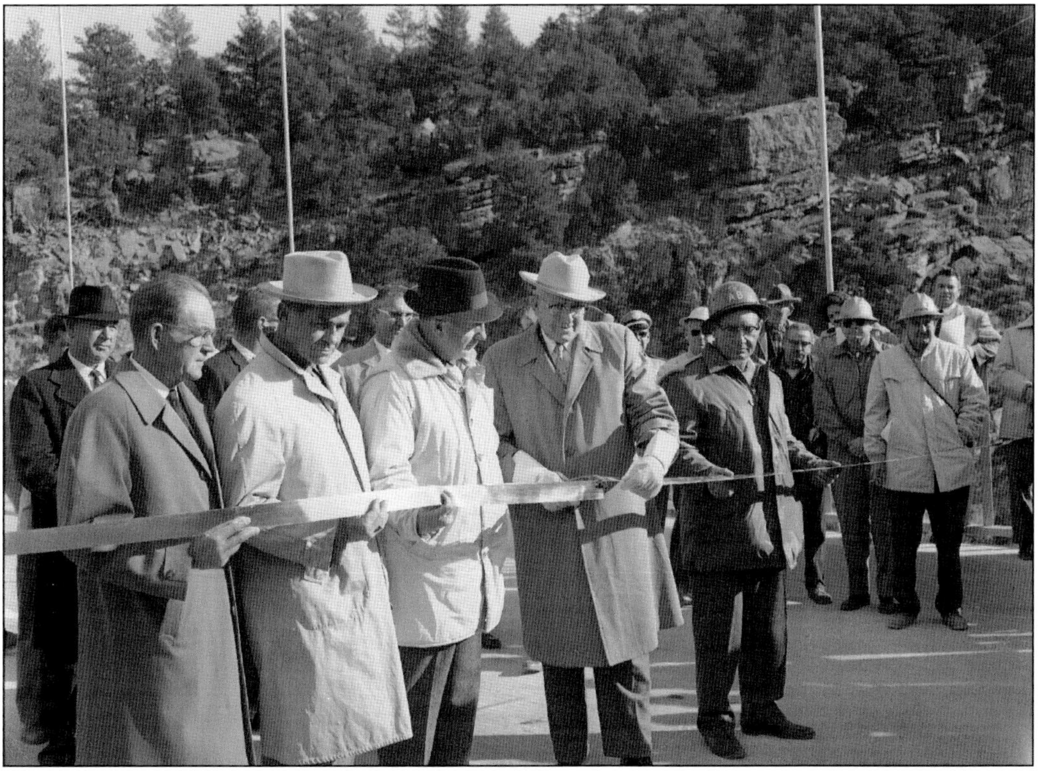

In the above photograph, Gov. George D. Clyde (center) officially shears a gleaming ribbon of steel at the dedication ceremony of the Cart Creek Bridge, signaling the span's completion. The governor spoke to the people, including officials from Utah and Wyoming, gathered in the northeastern corner of the state for the dedication. Featuring a number of engineering "firsts," the unusual bridge was designed by D. L. Sargent of the Utah State Highway Department. (Courtesy of UCRHC.)

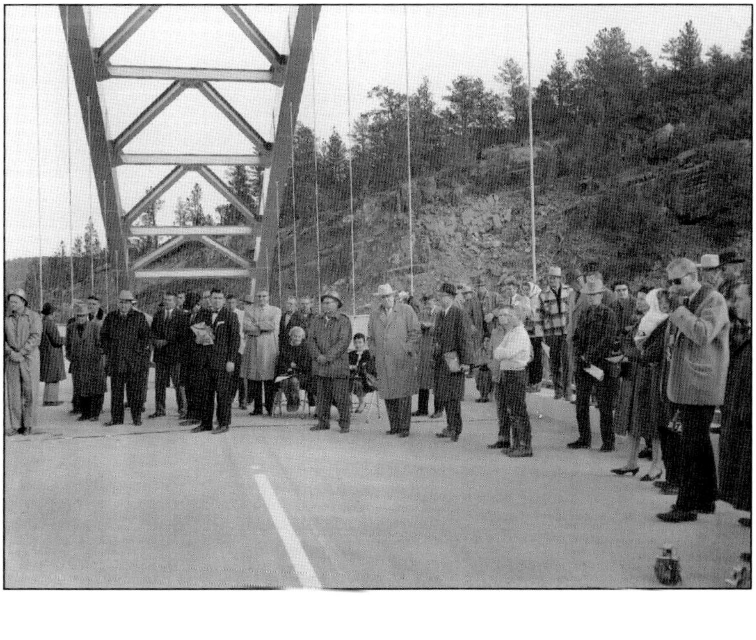

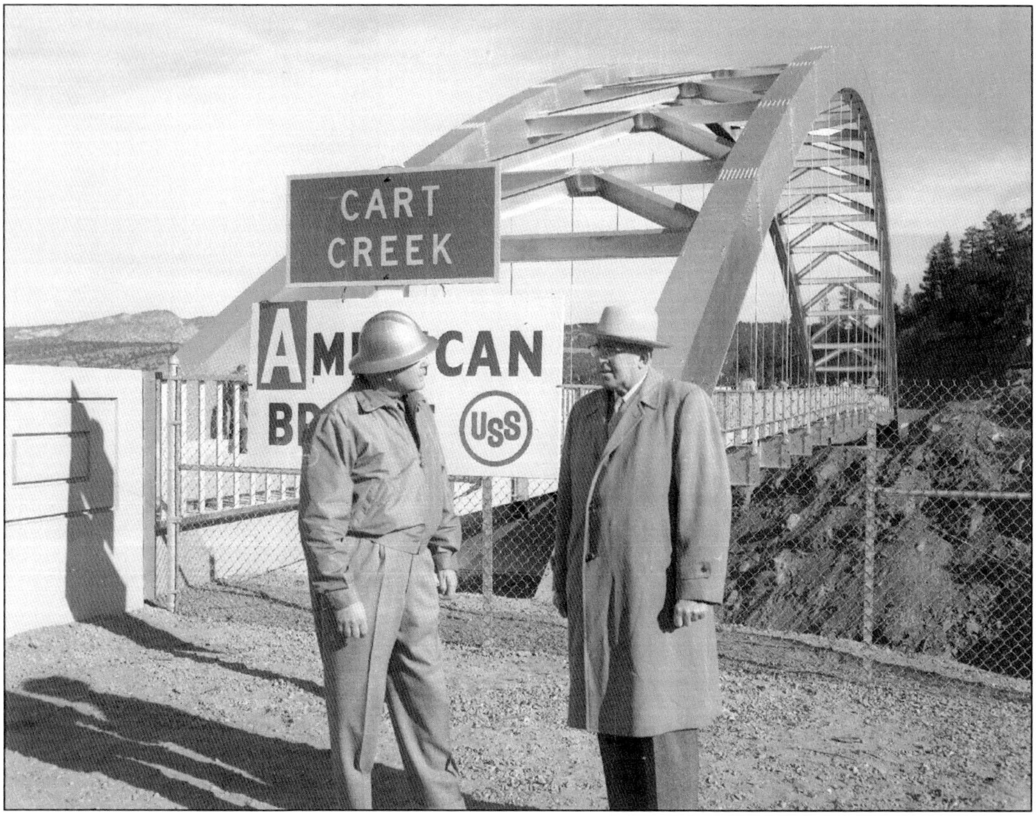

Jean R. Walton (left), Flaming Gorge Dam project construction engineer, speaks with Gov. George D. Clyde. As they visit, guests and dignitaries admire the longest single-span highway bridge with its beautiful rainbow arch. This bridge will be the link from Greendale Junction to the new town of Dutch John, Utah. (Courtesy of UCRHC.)

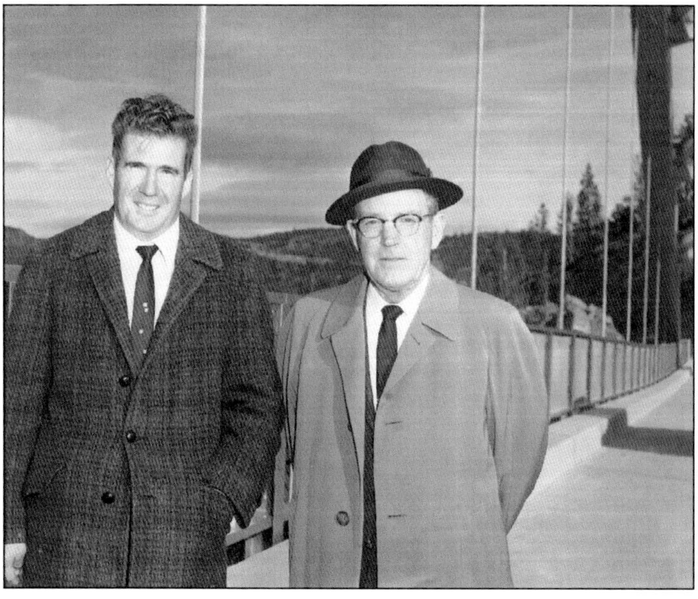

Joseph Burton (left), project engineer for the Utah State Department of Highway, and Francis Feltch, a member of the State Road Commission, are pleased with the work and completion of the award-winning Cart Creek Bridge. At the time of this photograph, traffic on this road terminated at the dam site because the dam was still under construction. The suspension bridge had been removed, and water had inundated the road to Dutch John from the north. (Courtesy of UCRHC.)

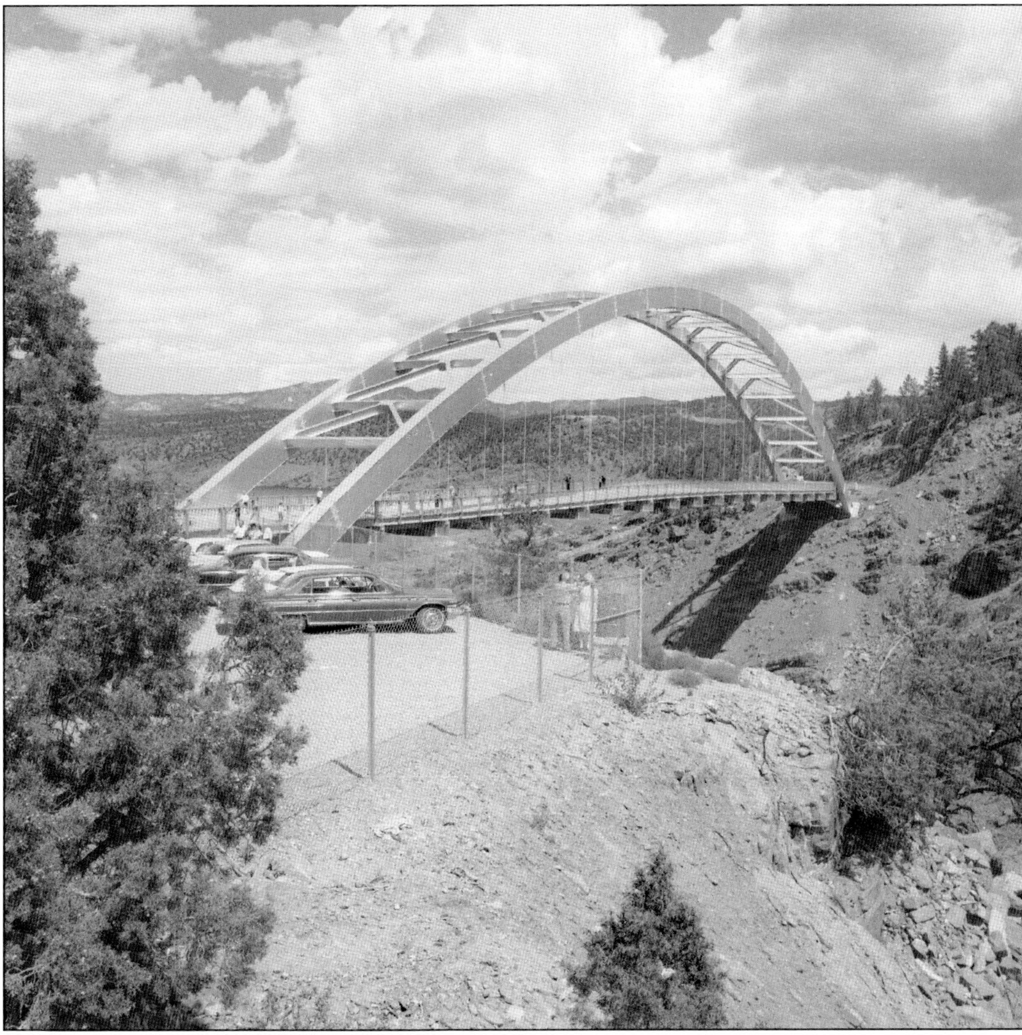

The contract for the Cart Creek Bridge was awarded to US Steel, American Bridge division, of Denver, Colorado, at a low bid of $768,782. The structure was the longest single-span bridge in Utah, at 550 feet. The bridge was a link to Flaming Gorge for the construction workers of the dam. The graceful bridge won an award of merit from the American Institute of Steel Construction for its beauty, and it served as a model for two other bridges built in Utah. (Courtesy of UCRHC.)

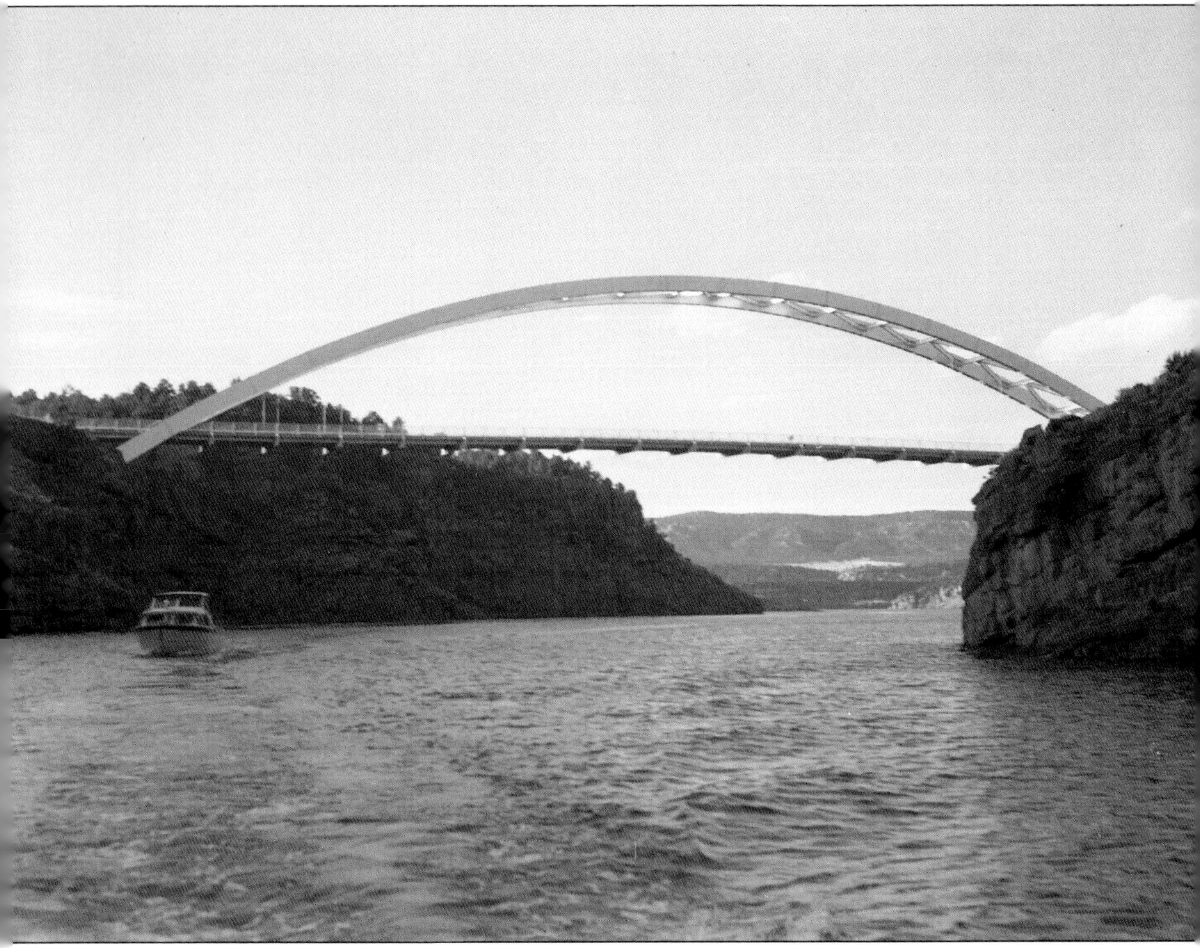

The construction of the Flaming Gorge Dam and Cart Creek Bridge are complete. The $810,000 bridge stands 200 feet above the floor of the rugged Cart Creek Canyon. This link between Vernal and Dutch John, Utah, is of great value as tourists and locals begin to visit the towns for shopping and recreation. The water of the Green River is backing up into the canyon to form the reservoir. The Cart Creek Bridge allows access over the canyon and to the dam site. The distance from the decking of the bridge to the water level, when filled, will be about 30 feet. It was a novelty for children to throw rocks from the bridge, but this was stopped because of the danger to boaters and fishermen below. (Courtesy of UCRHC.)

Six

Bucket by Bucket

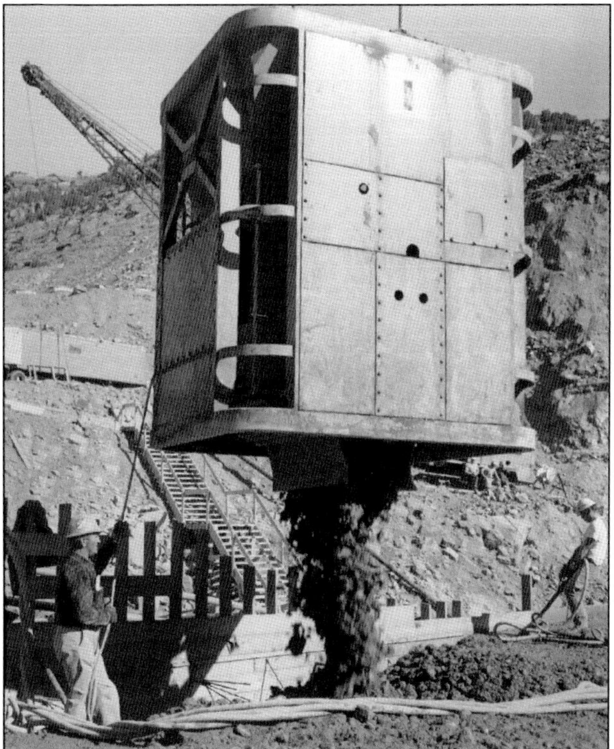

Harold Sitz, placement foreman for the dam, pulls the rope to release eight cubic yards of concrete into a section of the dam in October 1960. The bucket, shown here suspended from the cableway, discharges the concrete into the forms. (Courtesy of BOR.)

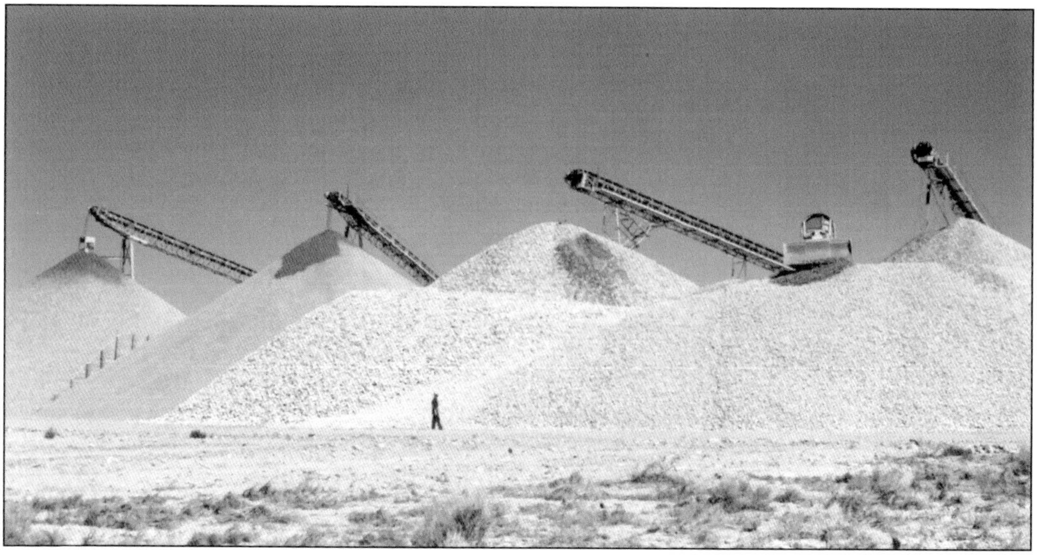

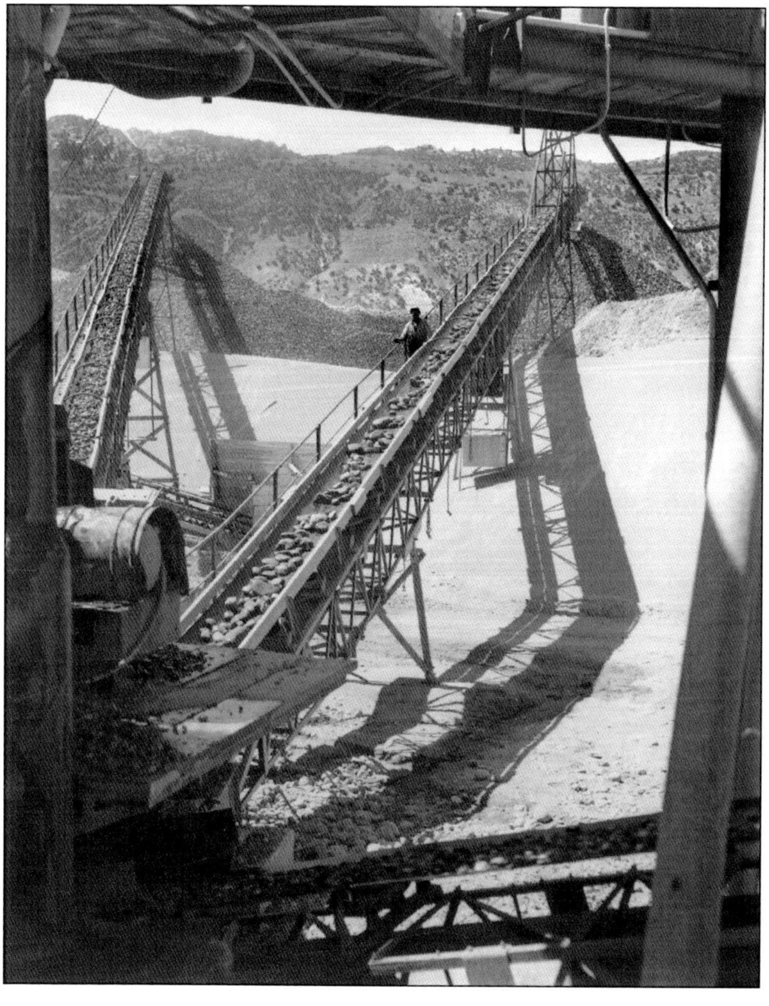

Above, stockpiles of aggregate are formed at the Henry's Fork Borrow on Williams Ranch near Linwood. The large plant provided gravel for the mixing plant at Flaming Gorge. The material was separated by size: three-fourths of an inch, an inch and a half, three inches, and six inches. It was carried to the proper stockpiles by conveyors, at left, which extended from the storage bins to the top of the plant. Large 18-wheelers would then haul the gravel from the aggregate plant to the batch plant at the dam. During the busy concrete-placing months, trucks were continually driving the 12-mile circuit. (Courtesy of BOR.)

The concrete batch plant houses four concrete mixers, four cubic yards in volume, which will produce the one million cubic yards of concrete required for the dam and power plant. The concrete is mixed in the batch plant, dumped into a transfer car, carried a short distance, and placed into the cubic-yard buckets. These will then be carried by the cableway system to the placement area at the dam or power plant. (Courtesy of BOR.)

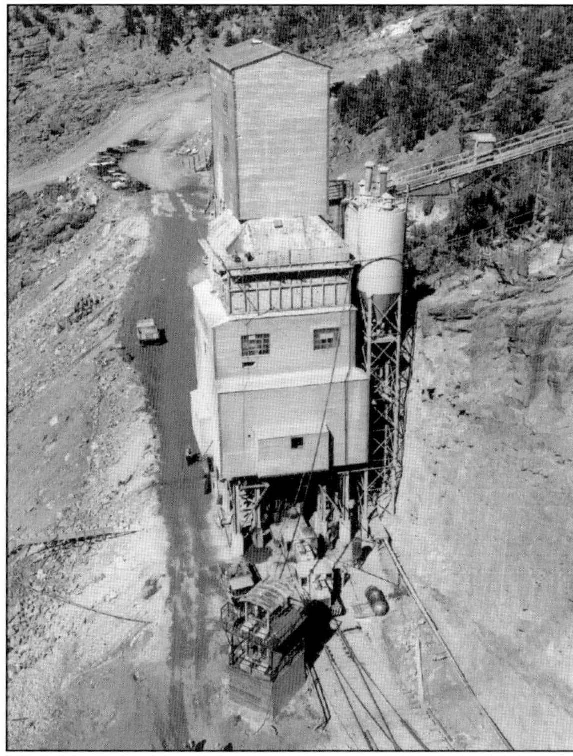

An electric-diesel transfer car is seen traveling a short distance from the concrete plant on Daggett County's only railroad, nicknamed "Flaming Gorge Short Line." The transfer car empties a batch of concrete into the eight-cubic-yard bucket, which will be pulled by way of cable system to the dam site, where it will be emptied. (Courtesy of BOR.)

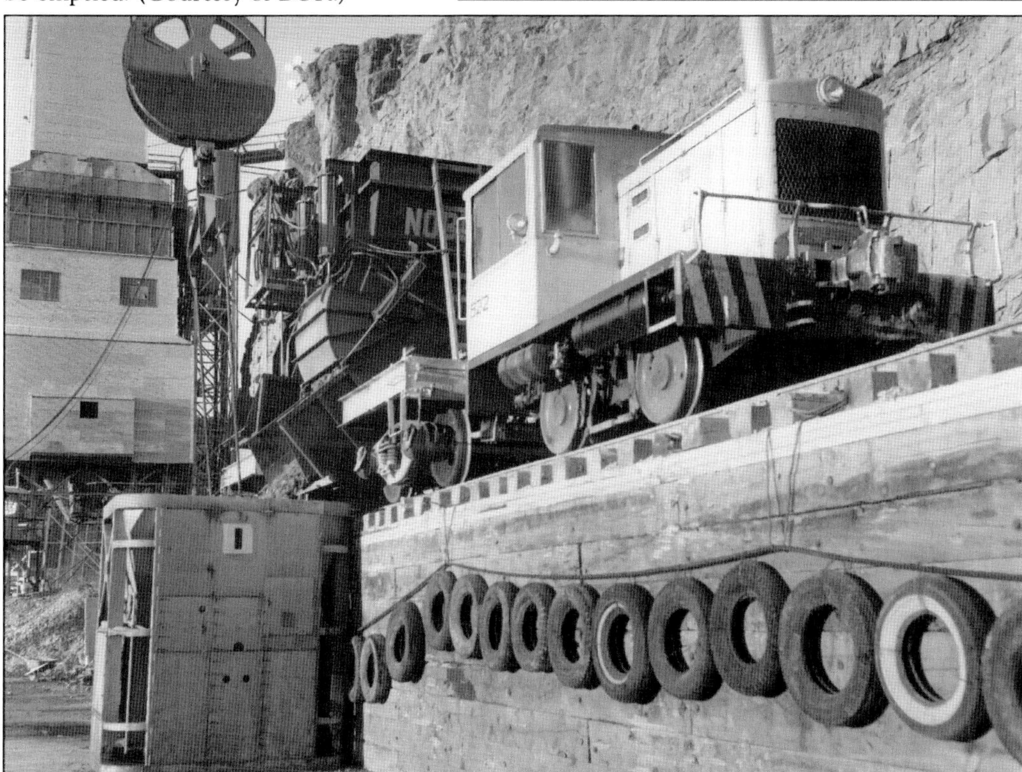

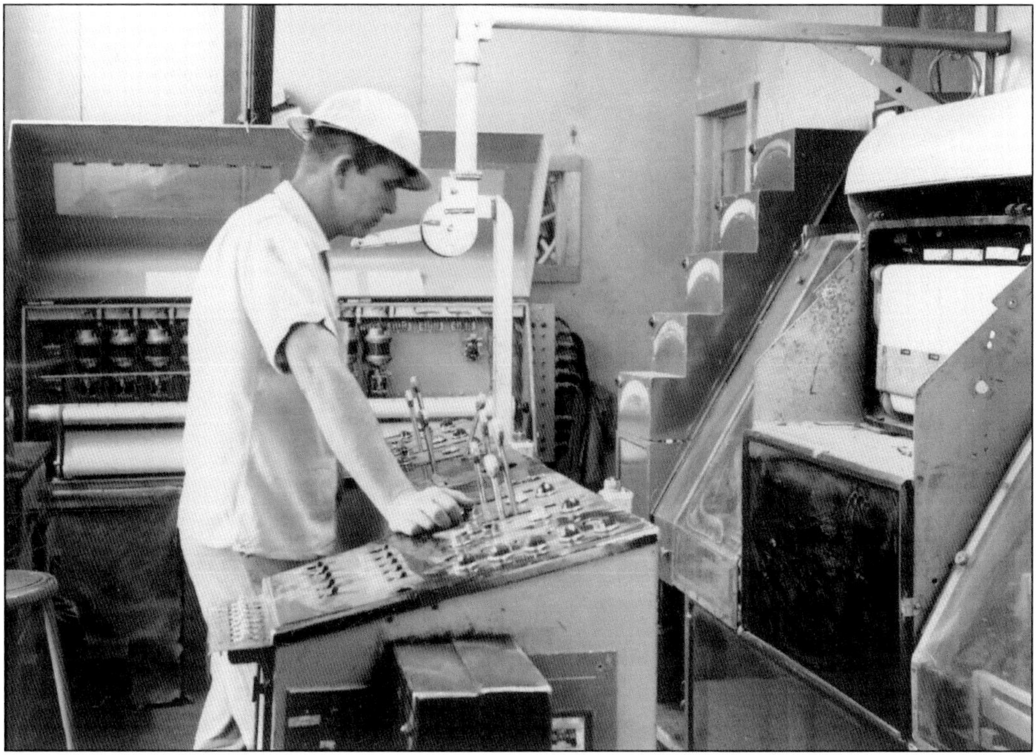

Frank Hines, the batch plant operator, manipulates the switches and levers on the control board, which is the nerve center of the large facility. Surrounding him are various weighing and recording machines that monitor all of the ingredients that go into each batch of concrete being mixed and placed in the dam. The photograph was taken on June 20, 1961. (Courtesy of BOR.)

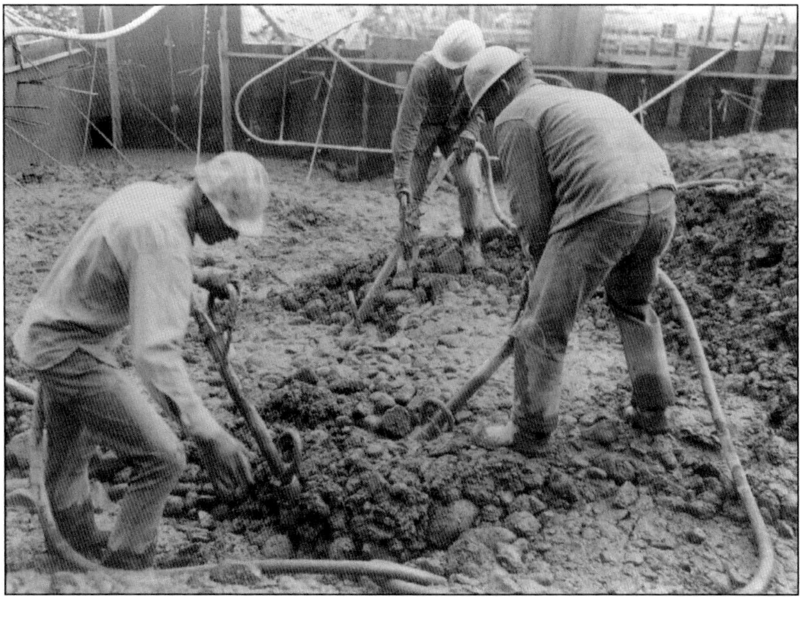

On October 13, 1960, vibrator operators work the large pile of concrete that has been dumped into the form on the dam. The concrete contains aggregate, rocks of up to six inches in diameter. The large vibrators were operated by compressed air. (Courtesy of BOR.)

A hive of activity is seen in this September 1960 overview of the construction of the Flaming Gorge Dam. Concrete is being placed in several blocks of the dam foundation, while forms are being built for additional blocks. The blocks were seven and a half feet thick. As each block was completed, it was sandblasted and washed before the next block was added on top. (Courtesy of BOR.)

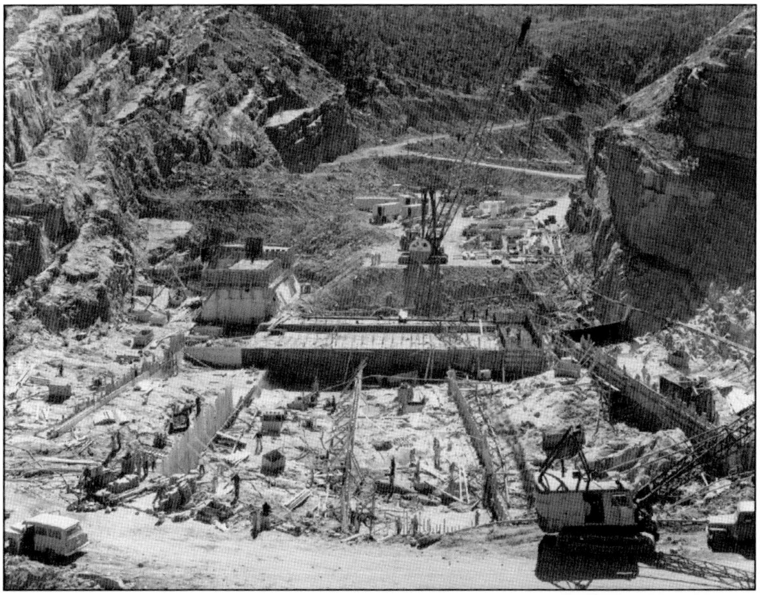

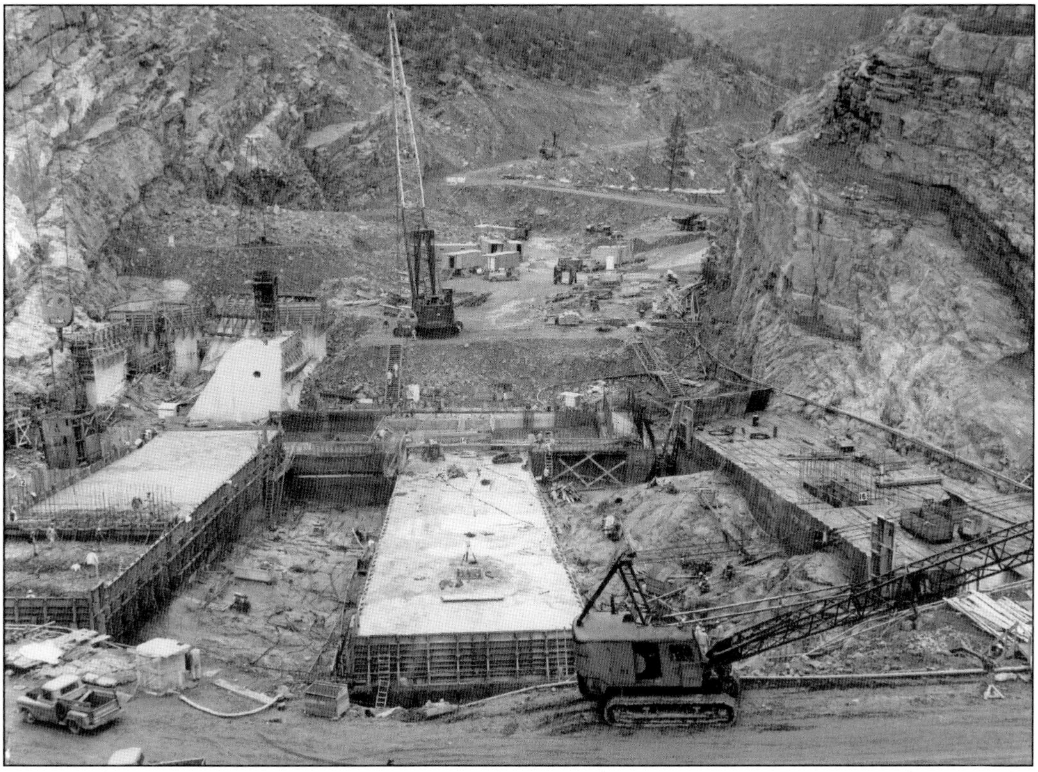

On October 13, 1960, a vast concentration of men and equipment can be seen working on the foundation of the dam. Several large placements of concrete are visible in the foreground. On the left, men are ready to receive another load of concrete from the bucket. Concrete has been placed in the substructure of the power plant, located at the downstream toe of the dam, seen in center behind the dam block sections. In the center left of the photograph are the concrete sections for the outlet works. (Courtesy of BOR.)

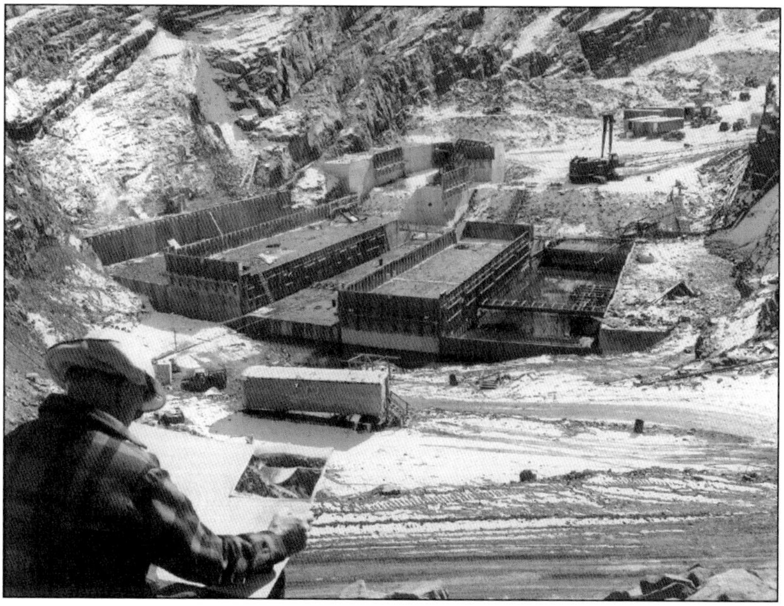

T.K. Moe, the safety engineer for the Flaming Gorge Dam project, attempts to anticipate future danger spots when concrete placing would begin following the winter months. This photograph is looking downstream in February 1961. (Courtesy of BOR.)

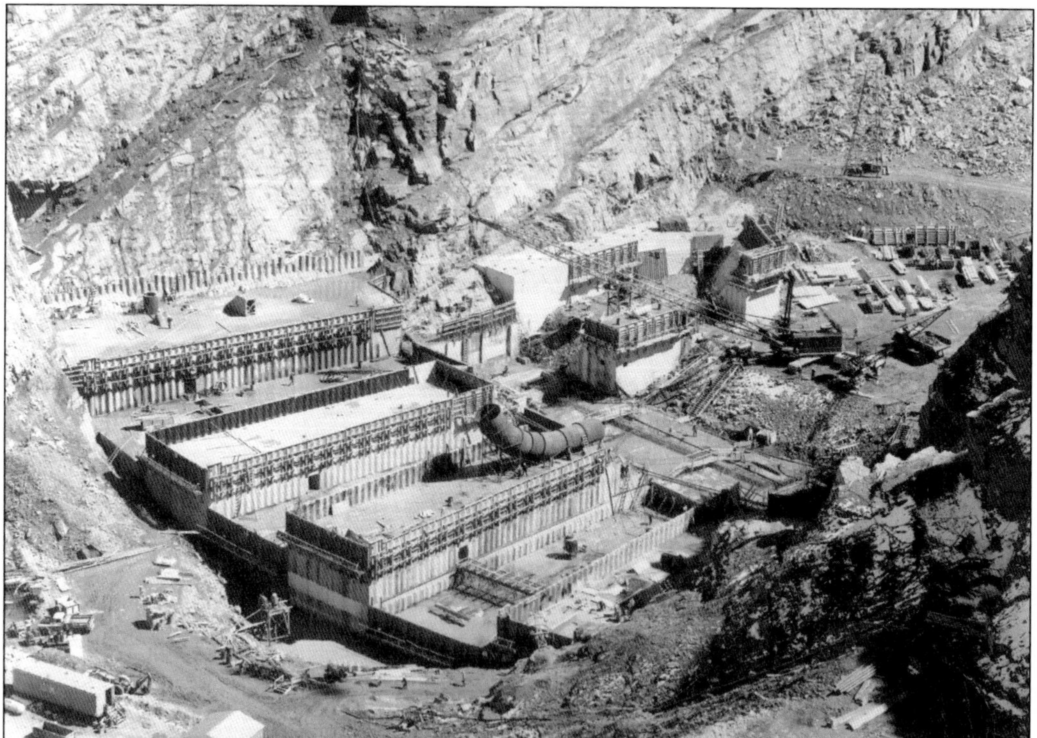

Work reaches a peak following a delay due to a snowstorm and freezing temperatures. During the next summer, 500 men will be employed on the dam project. In the center of this photograph is the first penstock (the culvert-shaped part), which will help to control the flow of water. The photograph was taken on March 30, 1961. (Courtesy of BOR.)

Prominent visitors view the progress on the Flaming Gorge Dam on April 28, 1961. Shown here are, from left to right, project engineer Jean R. Walton, Bureau of Reclamation commissioner Floyd E. Dominy, Secretary of the Interior Stewart L. Udall, and Bert Hanna from the *Denver Post*. (Courtesy of BOR.)

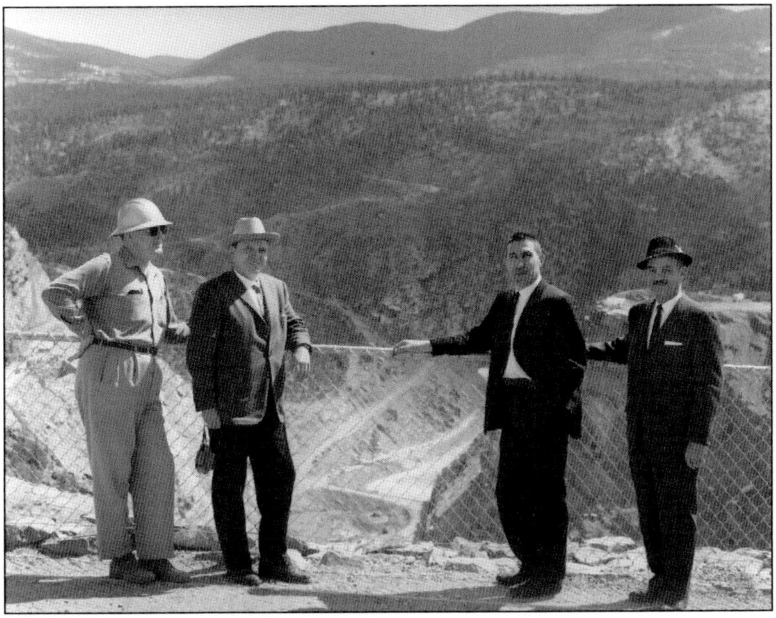

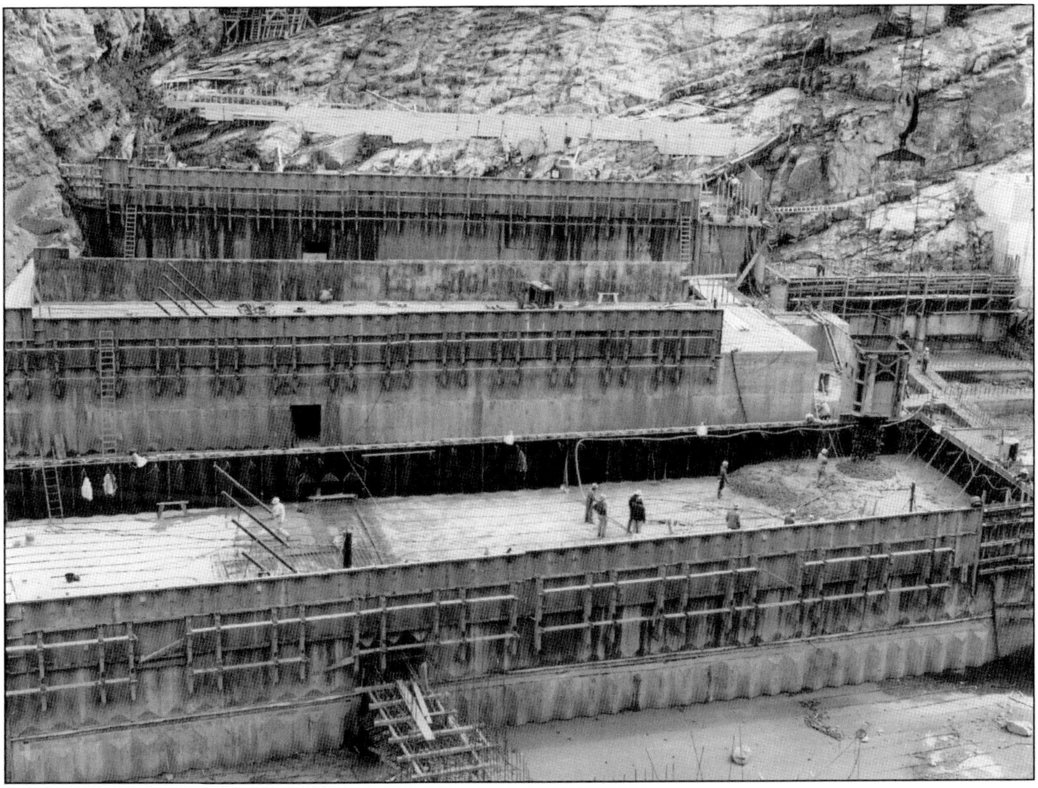

Shown here in March 1961, Flaming Gorge Dam begins to rise as concrete was steadily poured following the winter months. On the right, workmen, dwarfed by the growing structure, are busy pouring concrete. (Courtesy of BOR.)

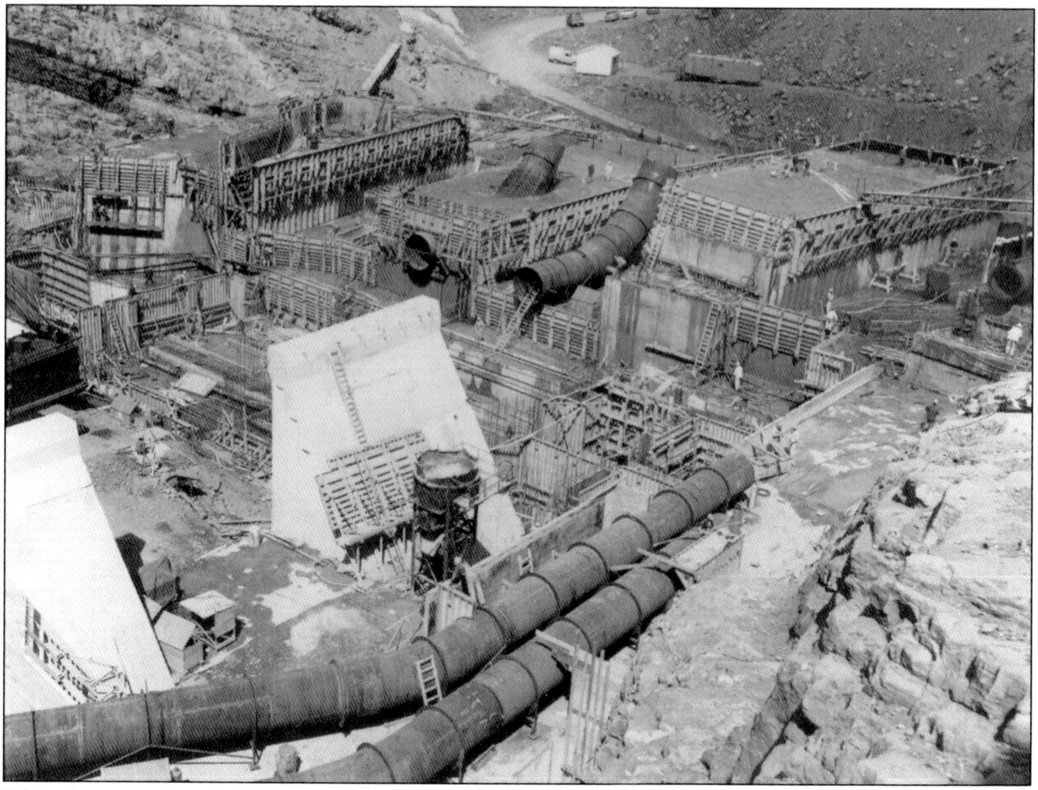

This photograph, taken on April 25, 1961, is looking upstream from the left abutment of the dam. The six-foot-diameter outlet pipes, seen in the foreground, will have an average length of 410 feet, with a maximum combined discharge of 4,000 cubic feet per second. The second of three penstocks is being embedded in a block of the dam. Each penstock is 10 feet in diameter and will rise 249 feet within the concrete dam structure. (Courtesy of BOR.)

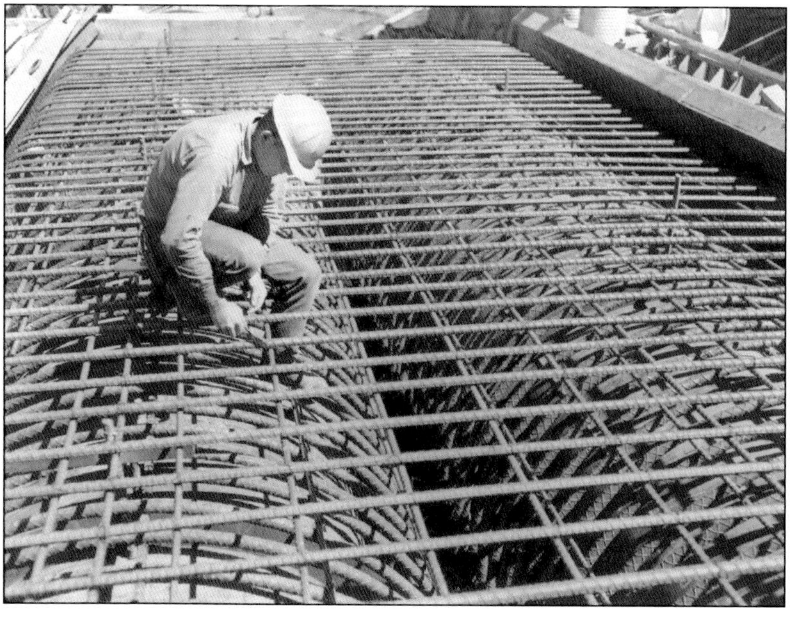

Joe Phelps, rod foreman, ties rods together, forming a reinforcement steel mat around the outlet pipes before they are embedded in concrete. This photograph was taken on May 22, 1961. (Courtesy of BOR.)

Eugene Boyt, the mechanical engineer for the Bureau of Reclamation, watches installation of the first penstock section for unit two in the powerhouse of Flaming Gorge Dam. There will be three penstock conduits, each conveying water to a 50,000-horsepower turbine. Each penstock is 10 feet in diameter and made of 1 1/16-inch high-grade steel. (Courtesy of BOR.)

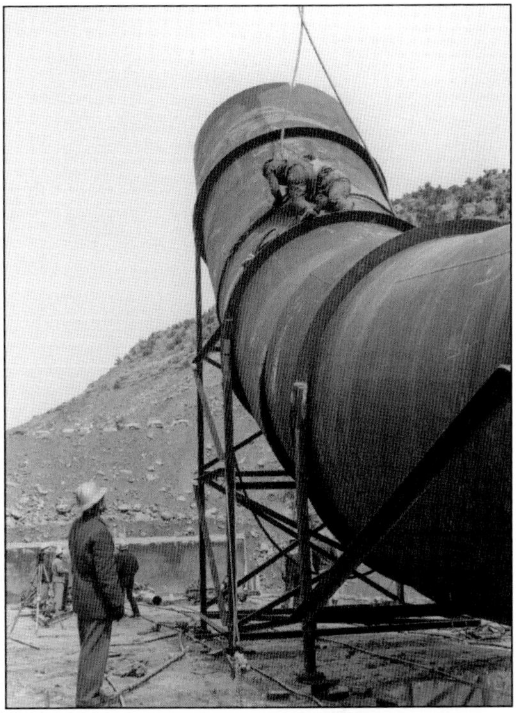

Work continues on a 24-hour basis on the Flaming Gorge Dam and powerhouse. The first sections of penstock for generating units one, two, and three in the powerhouse have been embedded in mass concrete. The two pipes for the river outlet are also seen buried in concrete. (Courtesy of BOR.)

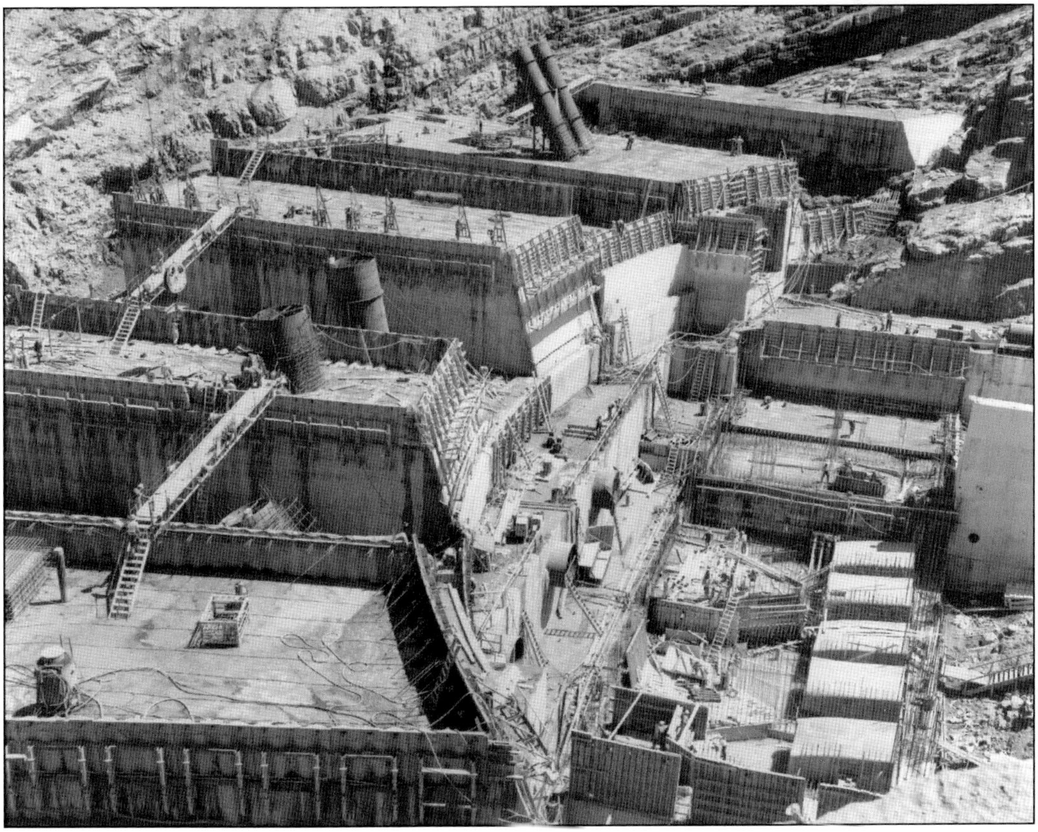

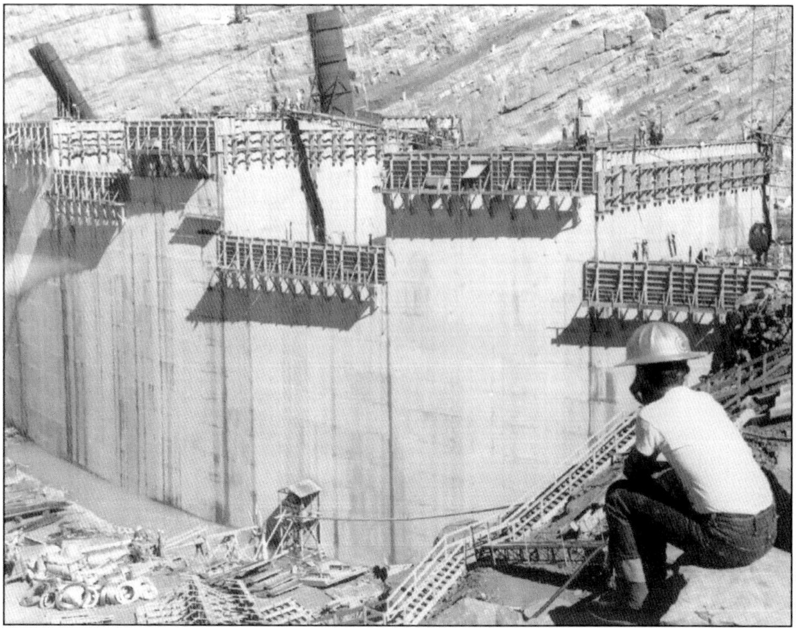

Craig Hammond, a 13-year-old from Vernal, watches the construction of the Flaming Gorge Dam and dreams of the day he may be an engineer of such a project. This photograph, taken on July 7, 1961, shows the reservoir side of the dam. (Courtesy of BOR.)

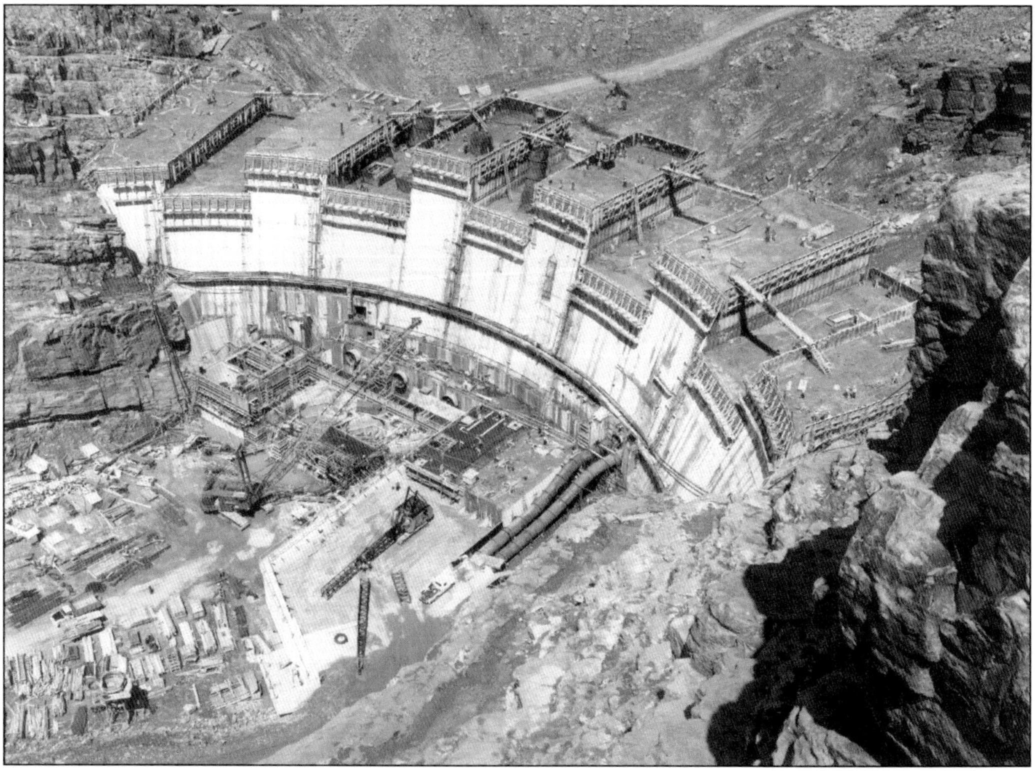

In this July 1961 photograph, taken during the peak of construction activity, workers can be seen swarming over the area like ants at a picnic. A quarter of a million yards of concrete has already been used in the dam by this time. An overall view of the construction has been obtained from the building on the lookout point near Dutch John. (Courtesy of BOR.)

Heading the construction of Flaming Gorge Dam for Arch Dam Constructors, primary contractors for the project, are project engineer Henry C. Scott (left), project manager Douglass D. Baker (center), and assistant project manager F.W. Bowman. This photograph was taken on September 5, 1961. (Courtesy of BOR.)

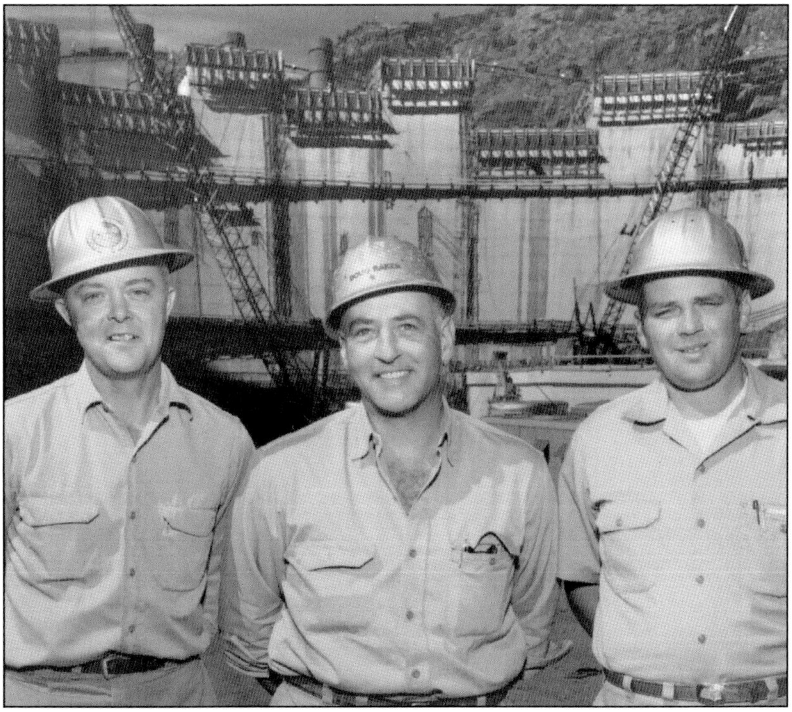

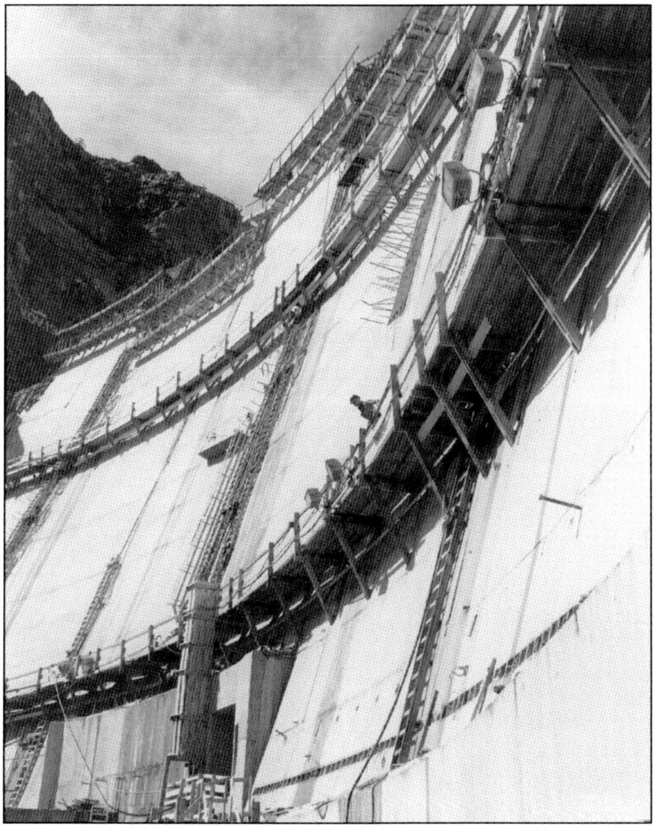

The size of a man on the catwalk offers a perspective as to the vastness of Flaming Gorge Dam. Mass concrete in the highest block is now 215 feet above the foundation. By September 14, 1961, when this image was taken, over a third of the one million cubic yards of concrete had been placed. (Courtesy of BOR.)

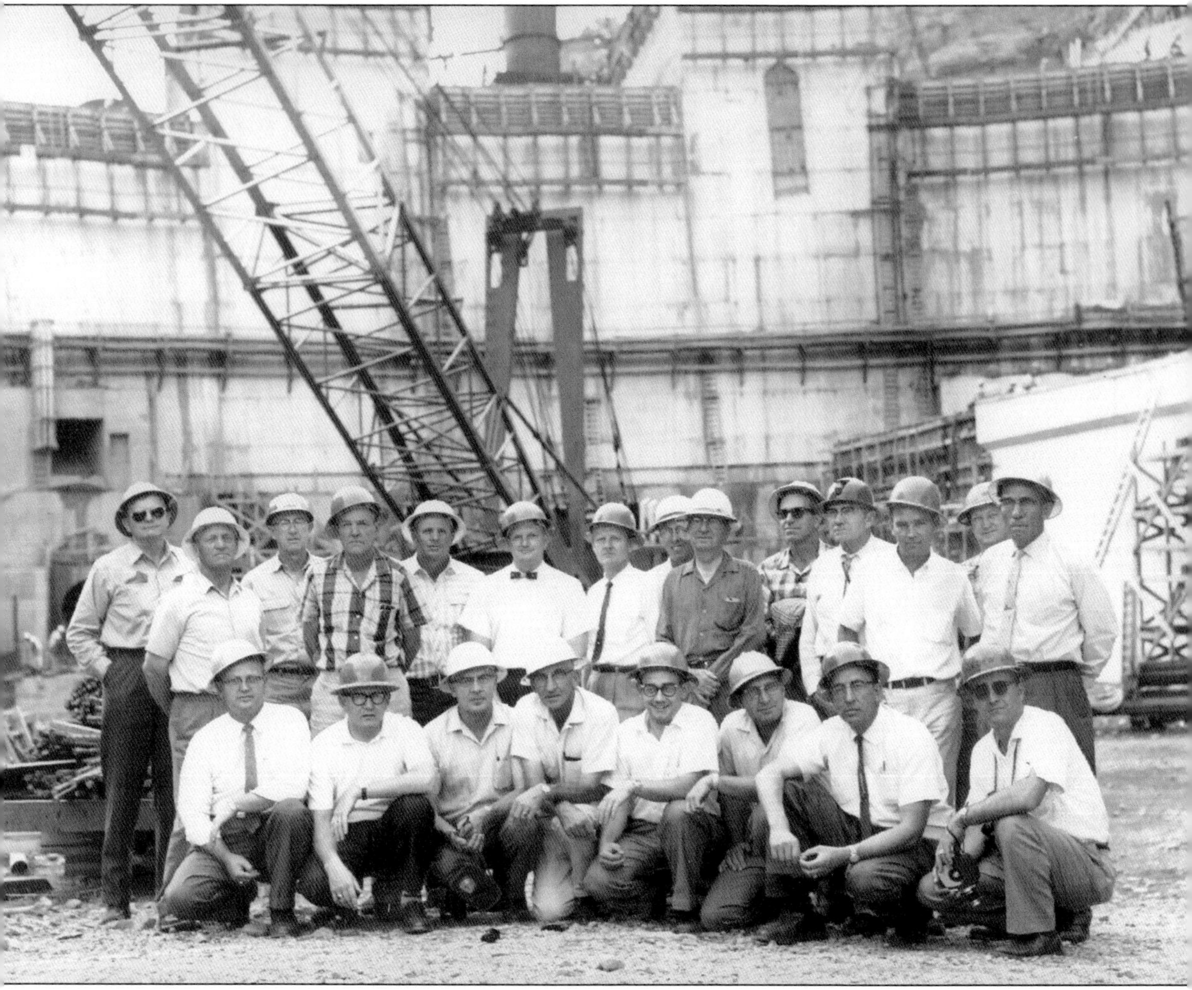

The Vernal Chamber of Commerce visited the Flaming Gorge Dam construction site on July 19, 1961. They are shown here with members of the Bureau of Reclamation. Pictured are, from left to right, (first row) J.C. Haws, Barclay Gardner, L.Y. Siddoway, Commissioner Rulon S. Hacking, chamber president Paul M. Armstrong, Royal Henderson, Hal Anderson, and Dr. R. Bruck Christian; (second row) construction engineer Jean R. Walton, county commissioner Harold McKee, unidentified bureau official, Francis Feltch, Roland Merkley, Rex Roylance, unidentified bureau official, Harold Duke, B.H. Stringham, Dr. T.R. Seager, C.J. Neal, D.L. Wetencamp, unidentified bureau official, and Buell Bennett. (Courtesy of BOR.)

In August 1961, Donell Carlson, the five-year-old daughter of Mr. and Mrs. Don Carlson, is silhouetted against the window of the lookout house that faces the Flaming Gorge Dam construction. The upstream side of the dam can be seen, with a cofferdam between the dam and the diversion tunnel, where water is bypassed through the mountain during construction. (Courtesy of UCRHC.)

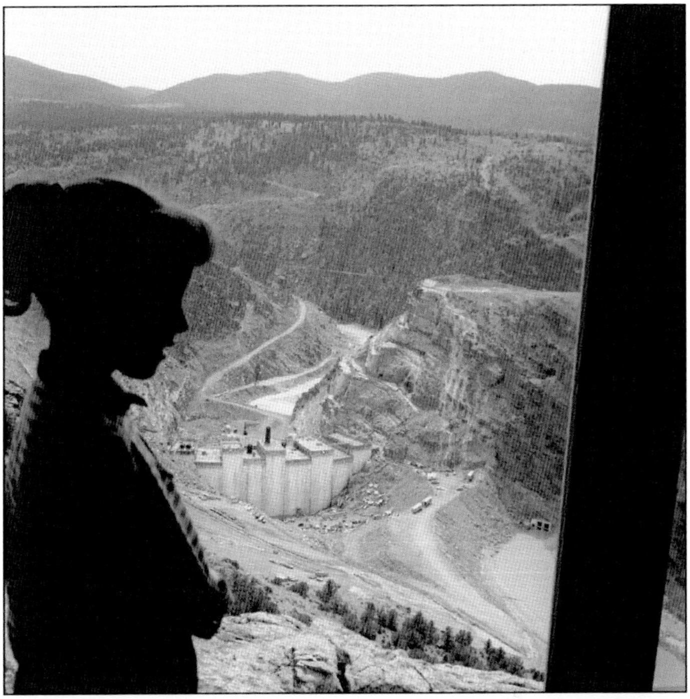

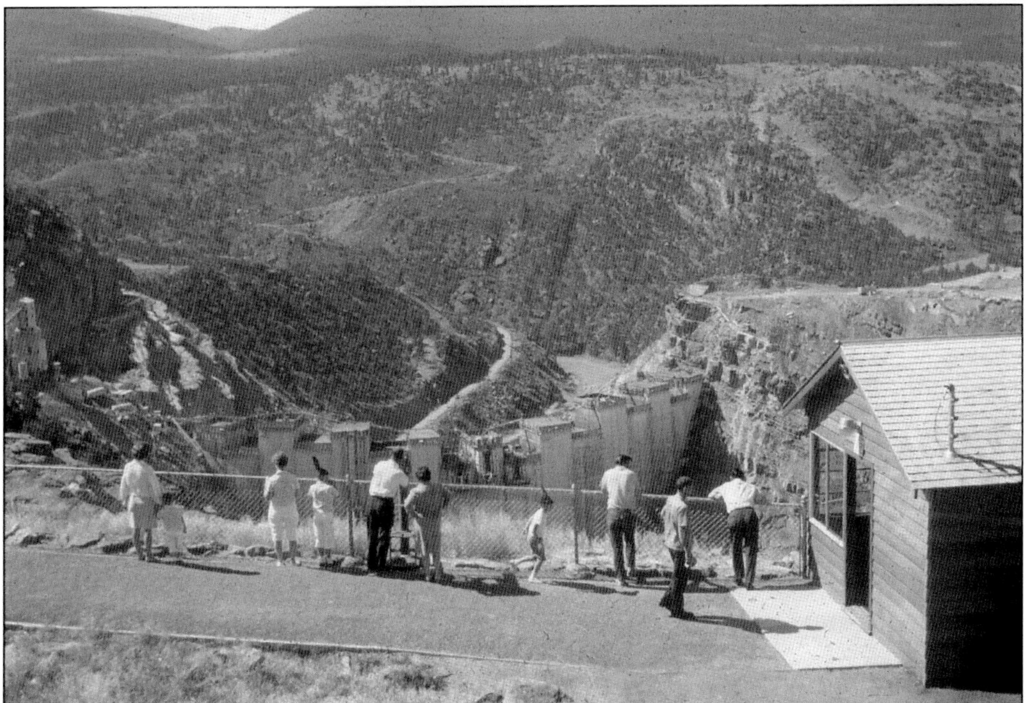

In June 1962, visitors view the Flaming Gorge Dam construction from a vantage point northeast of the dam. A telescope mounted near the fence offers visitors a close-up of the construction. From this same spot, one can see the famed Ashley Falls and parts of the deep Red Canyon. (Courtesy of USFS.)

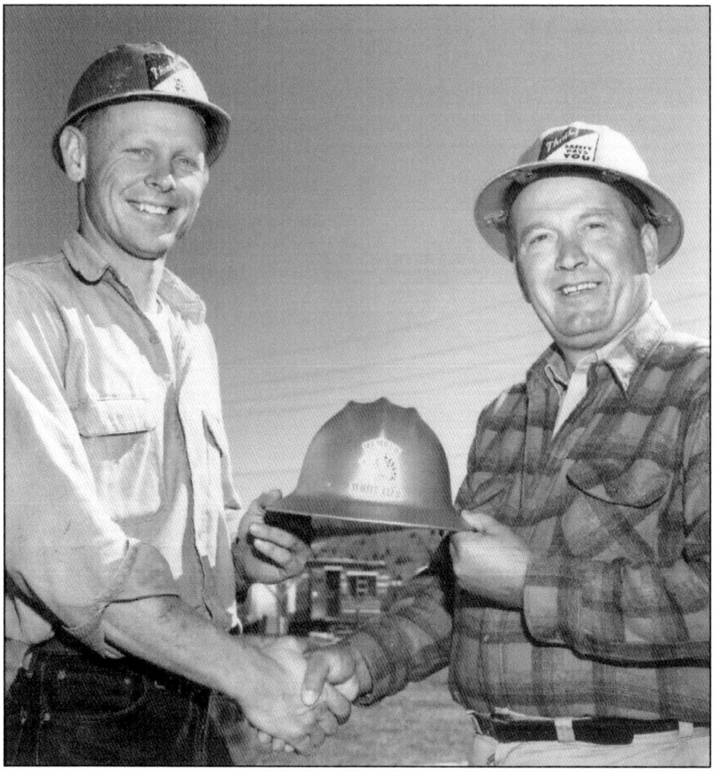

Harold B. Oney (left), of Dutch John, Utah, was presented with a pin, certificate, and hard hat by Maynard Jean, safety engineer for the Flaming Gorge Dam project, on September 27, 1961. Oney, a carpenter, was saved from a serious head injury from a falling rock because he was wearing his hard hat. He became the sixth member of the Turtle Club. Membership in the club is attained by being spared a fatal head injury from use of a hard hat. (Courtesy of BOR.)

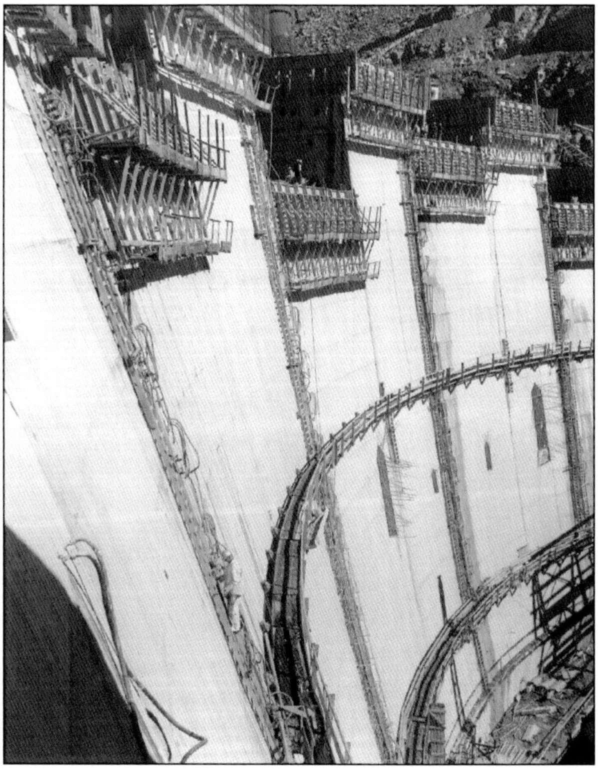

Dave Sathers, of Craig, Colorado, works from a ladder in the lower left in this September 1961 photograph. Workmen's relative size diminishes as the dam increases in size. Concrete has been placed in 15 blocks, reaching a height of 230 feet above the foundation. (Courtesy of BOR.)

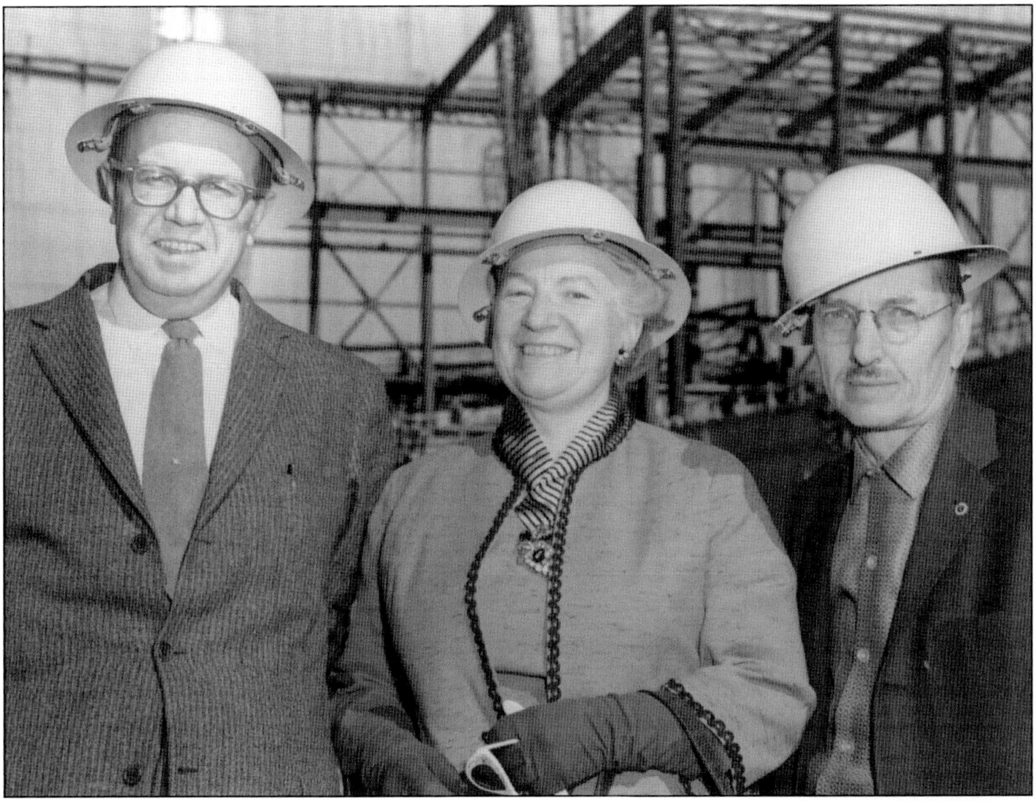

Wyoming dignitaries visited the construction site on October 13, 1961. Shown here are Sen. J.J. Hickey (left), Velma Linford, superintendent of public instructions for the State of Wyoming (center), and Adrian Reynolds, editor of the *Green River Star* and representative in the Wyoming Legislature from Sweetwater County. (Courtesy of BOR.)

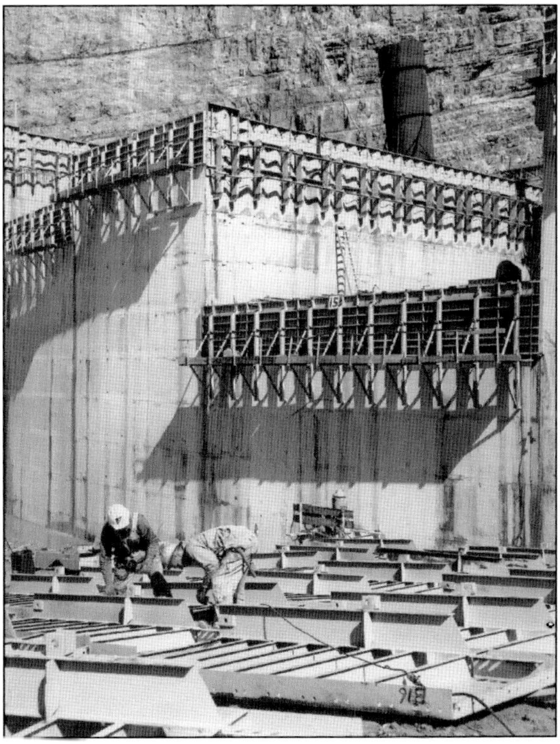

On June 8, 1961, Dellos W. Morrell (left), of Pleasant Grove, Utah, and R.D. Blackett, of Ogden, Utah, assemble steel cantilever forms for concrete placement on the dam's upstream face. The large pipe in the upper right is a section of the penstock that will convey water to the 50,000-horsepower turbines. (Courtesy of BOR.)

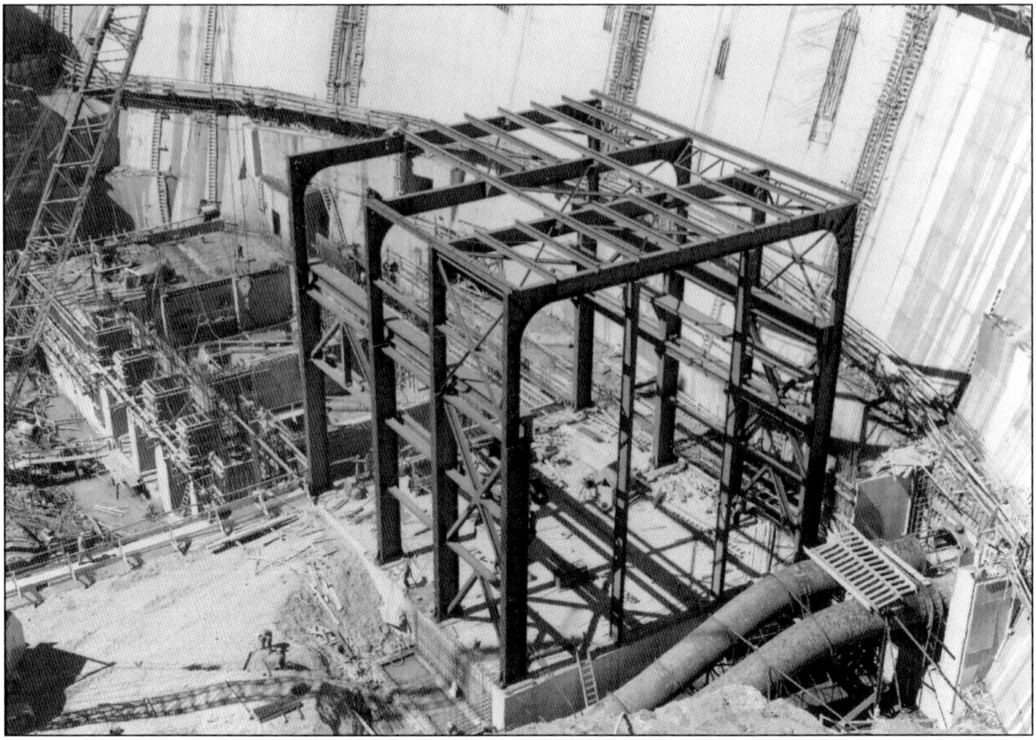

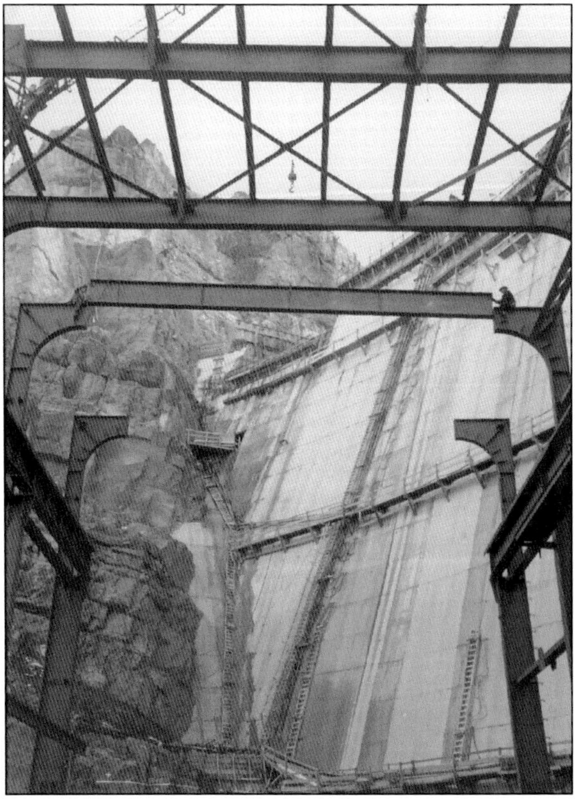

The first structural steel framework rises up on the powerhouse. It will contain three 36,000-kilowatt generating units. A section of the six-foot-diameter river outlet pipes are seen on the right. (Courtesy of BOR.)

Ironworkers, one on each end, place in position one of the massive girders in the Flaming Gorge Dam power plant. The tremendous weight required very careful handling to prevent accident to both the structure and the workmen. This photograph was taken on October 30, 1961. (Courtesy of BOR.)

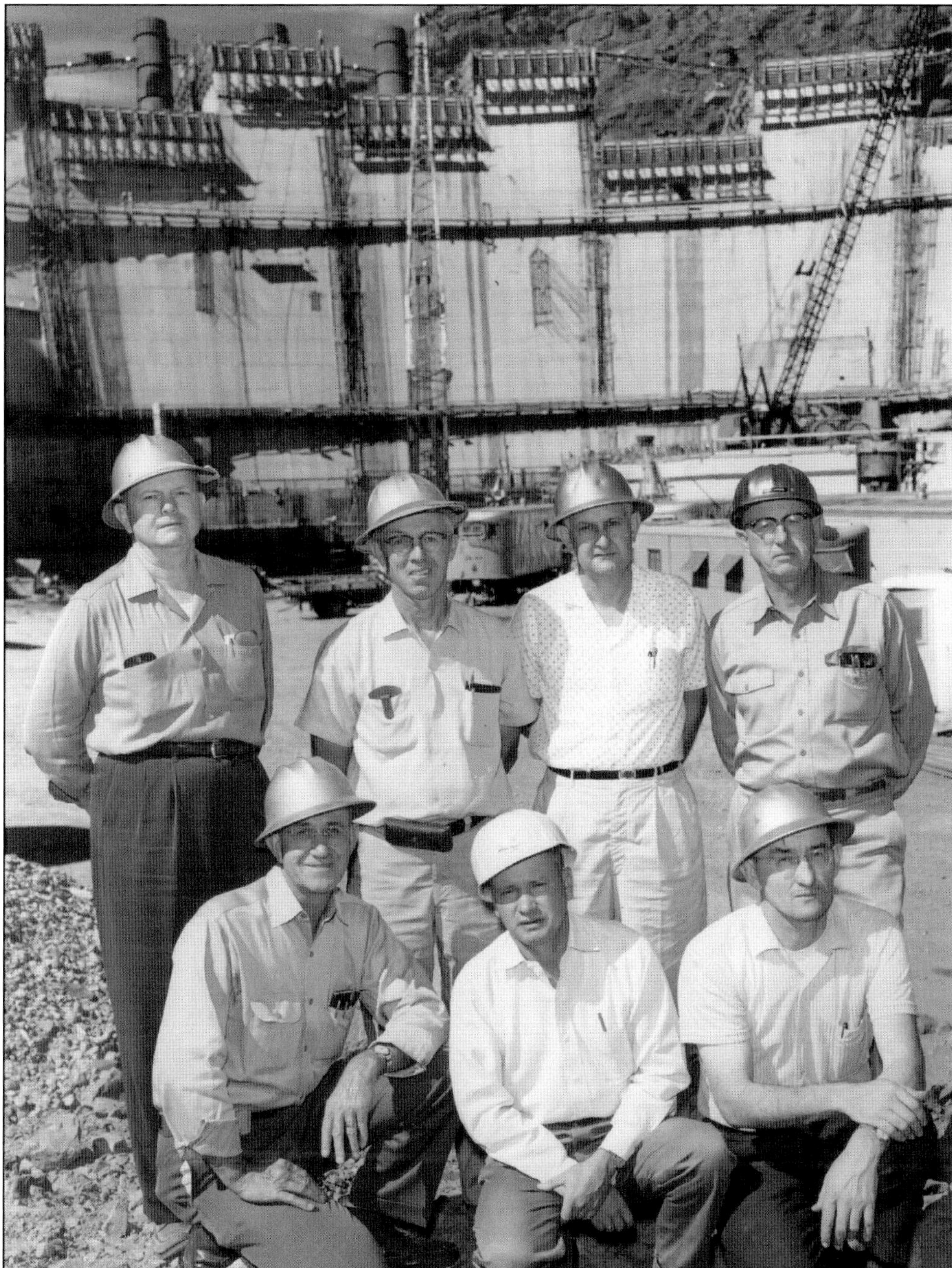

Project construction engineer and division chiefs pose in front of the dam construction on September 5, 1961. Shown here are, from left to right, (first row) Theodore W. Grover (chief of transmission line field division), Eugene P. Boyt (chief of mechanical and electrical engineering division), and Wesley A. Behling (chief of office engineering division); (second row) Jean R. Walton (project construction engineer), Frank S. Dallon (assistant project engineer), George Hensley Jr. (chief of administrative services division), and Russell C. Borden (chief of civil engineering division). (Courtesy of BOR.)

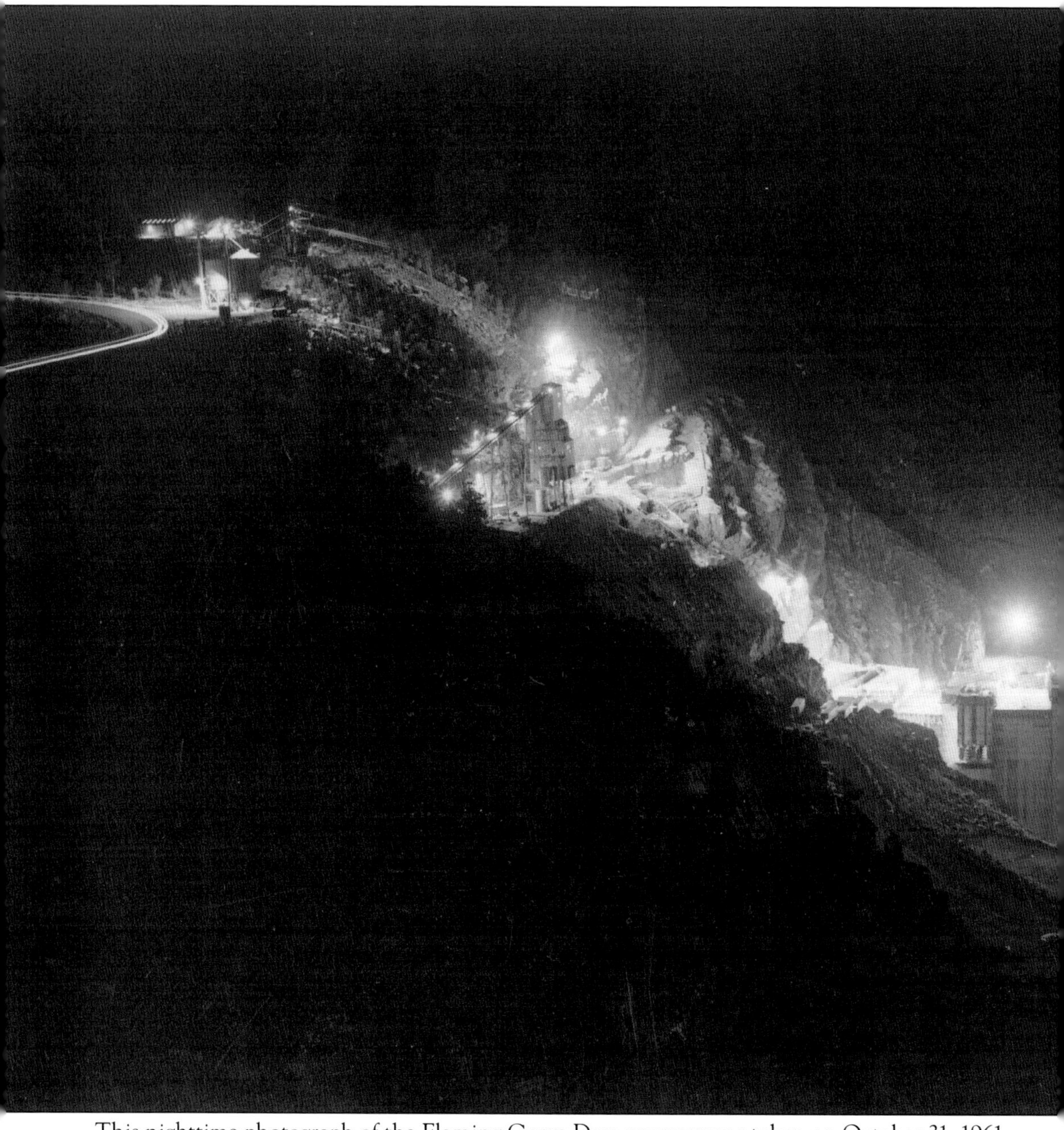

This nighttime photograph of the Flaming Gorge Dam progress was taken on October 31, 1961. Looking downstream reveals a beautiful panoramic view. On the left are the hoist house for the cableways and the storage silos for the cement. On the far right are the two tail towers that are

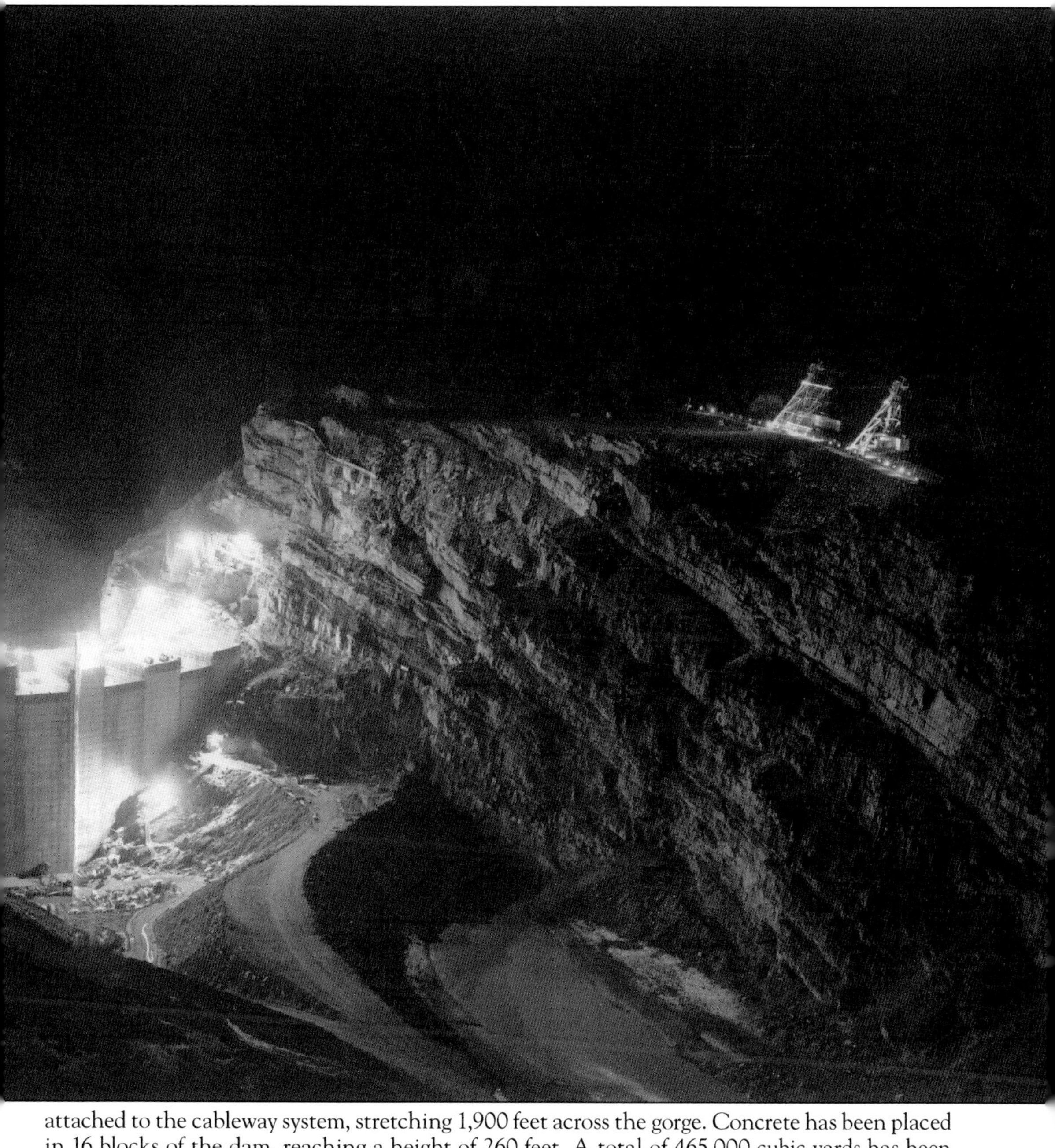

attached to the cableway system, stretching 1,900 feet across the gorge. Concrete has been placed in 16 blocks of the dam, reaching a height of 260 feet. A total of 465,000 cubic yards has been placed at the time of this photograph. (Courtesy of BOR.)

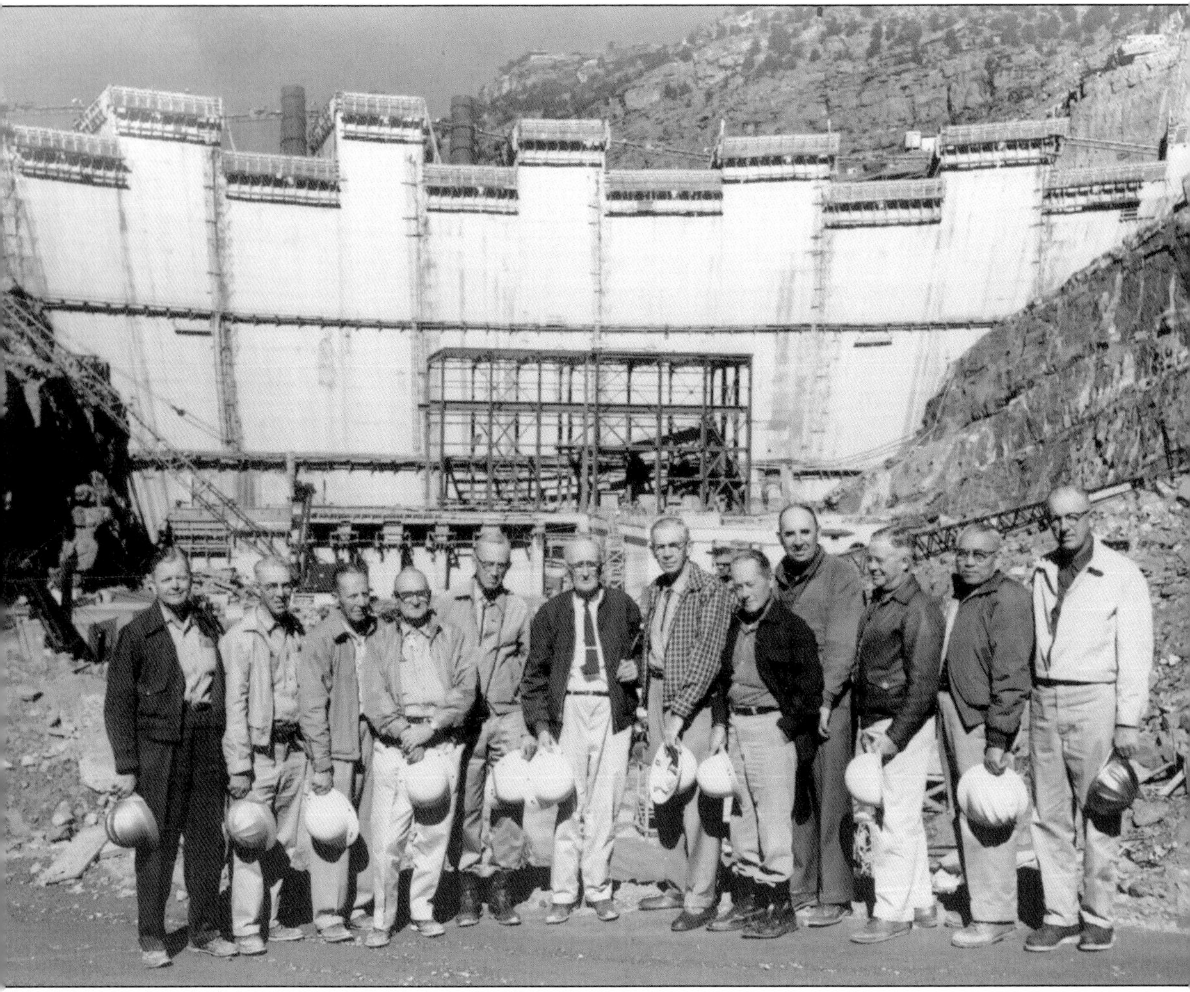

In October 1961, a board of consultants visited the dam and power plant project to inspect and review the construction and design. Shown here are, from left to right, Flaming Gorge Dam project engineer Jean R. Walton, assistant project engineer Frank S. Dallon, chief designing engineer O.L. Rice, consulting engineers Dr. Raymond E. Davis and John J. Hammond, consulting engineer chairman Julian Hinds, consulting engineers Dr. John W. Vanderwilt and Edward B. Burwell Jr., chief of Concrete Dam Designs E.R. Schultz, assistant regional director C.H. Carter, project geologist George Wonwai, and chief of Civil Engineering Division Russell Borden. (Courtesy of UCRHC.)

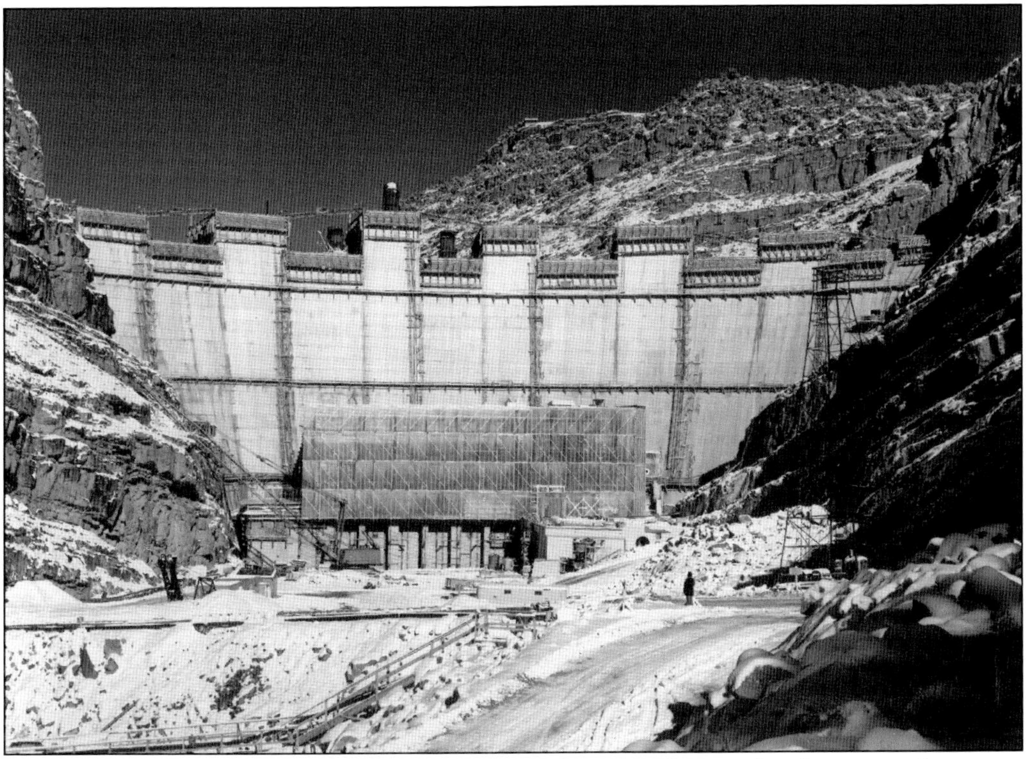

Winter cold and snow in the dam construction area brought a halt to the placement of concrete during the winter months. This December 7, 1961, photograph is looking upstream from the lower cofferdam. The powerhouse is being winterized with the erection of visqueen panels, allowing work to continue in that section of the project. During 1961, almost half of the total concrete was placed, reaching 260 feet in the highest block. (Courtesy of BOR.)

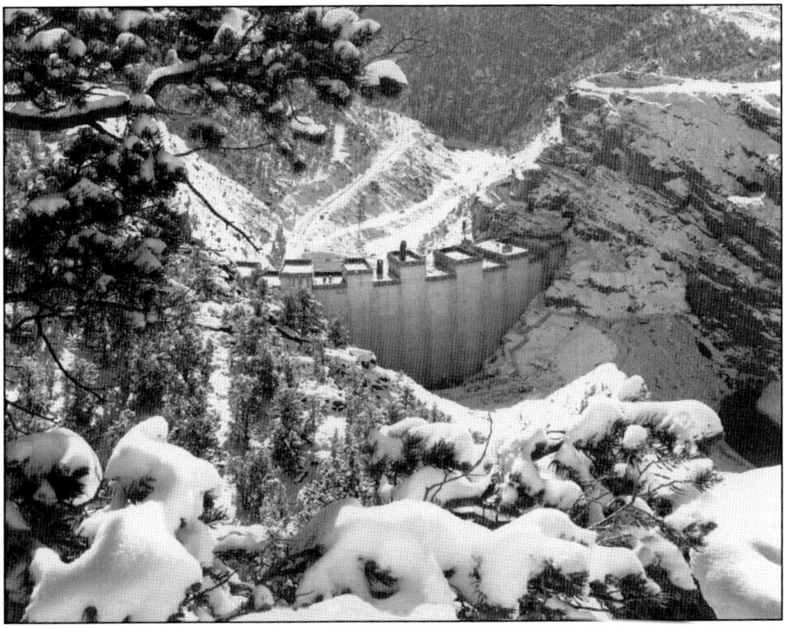

The Flaming Gorge Dam project is framed in the snow-covered boughs of a cedar tree. The photograph was taken looking downstream from the visitor's viewpoint in December 1961. (Courtesy of BOR.)

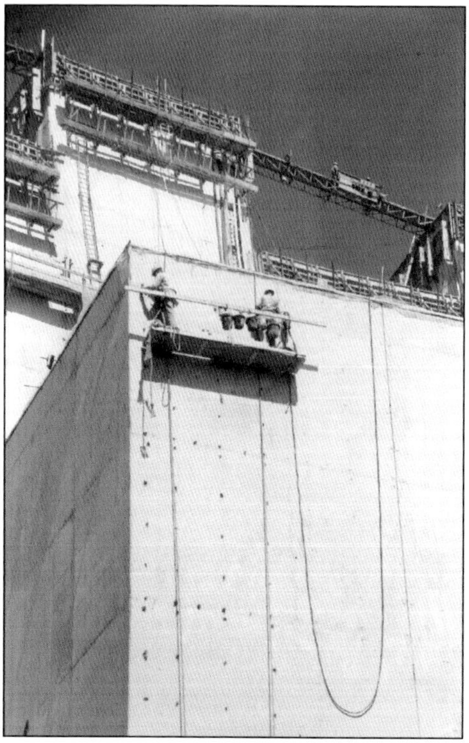

Looking up at the downstream face of the power plant, men on scaffolding are seen cementing in bolt-holes on March 29, 1962. Some of the high blocks of Flaming Gorge Dam loom in the background. (Courtesy of BOR.)

The placement of concrete is in full swing on the Flaming Gorge Dam following the winter months. The highest block is now 275 feet above the foundation. Progress is also rapid on the powerhouse, nestled in front of the dam. The dam and reservoir will be the heart of a new scenic and recreational wonderland. This photograph was taken on April 11, 1962. (Courtesy of BOR.)

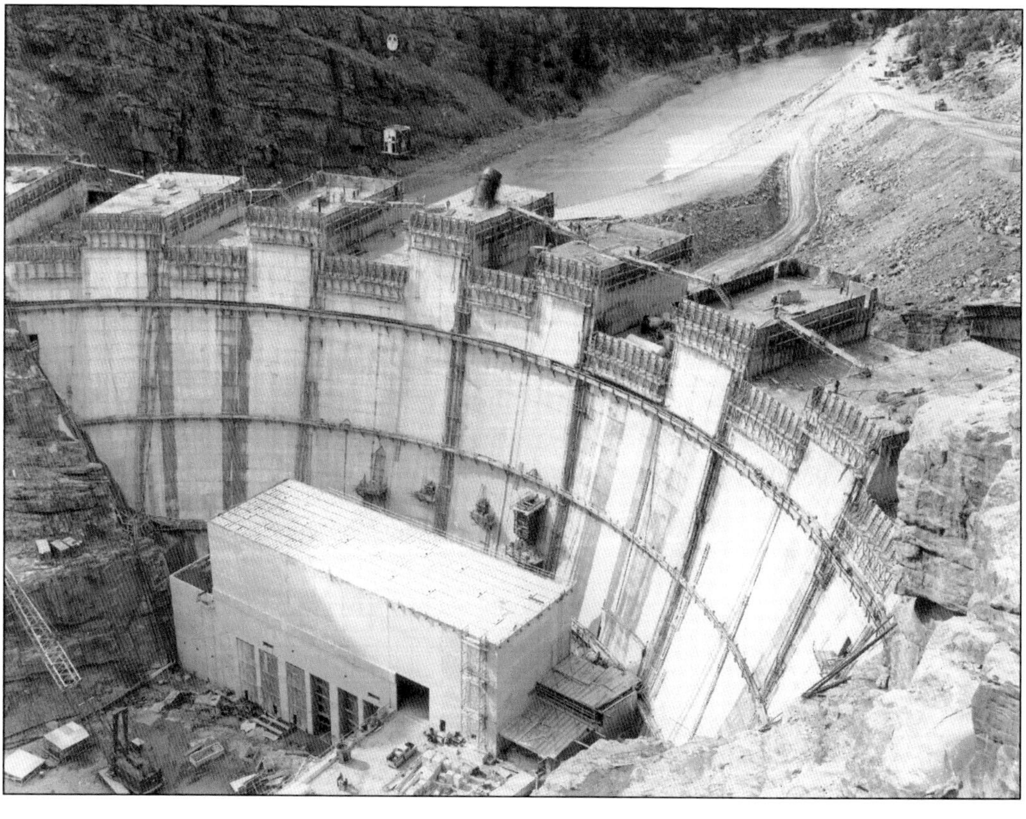

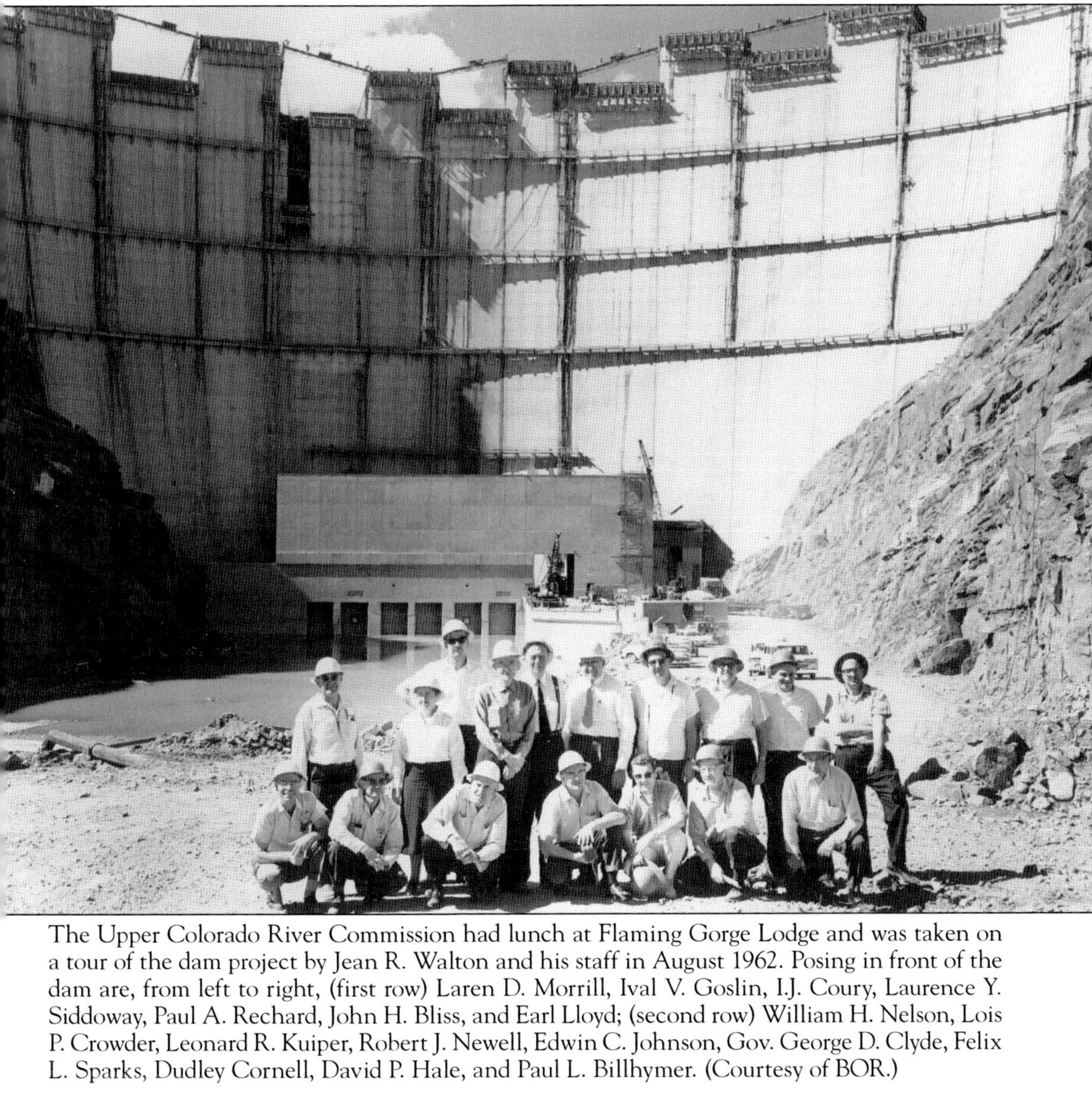

The Upper Colorado River Commission had lunch at Flaming Gorge Lodge and was taken on a tour of the dam project by Jean R. Walton and his staff in August 1962. Posing in front of the dam are, from left to right, (first row) Laren D. Morrill, Ival V. Goslin, I.J. Coury, Laurence Y. Siddoway, Paul A. Rechard, John H. Bliss, and Earl Lloyd; (second row) William H. Nelson, Lois P. Crowder, Leonard R. Kuiper, Robert J. Newell, Edwin C. Johnson, Gov. George D. Clyde, Felix L. Sparks, Dudley Cornell, David P. Hale, and Paul L. Billhymer. (Courtesy of BOR.)

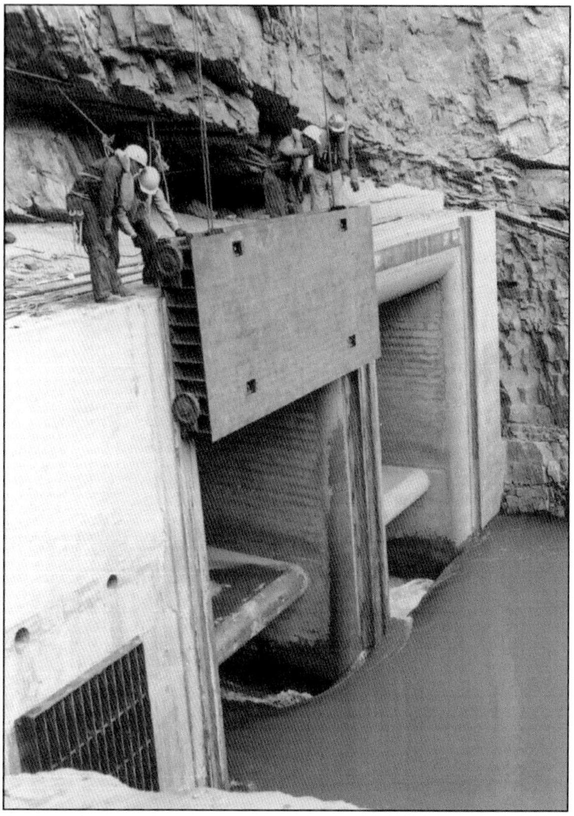

On November 1, 1962, workmen maneuver steel gates over the opening of the bypass tunnel at Flaming Gorge Dam. The newly created reservoir began rising steadily following the closing of the diversion tunnel. A portion of the tunnel remained open for a month before being completely closed off. (Courtesy of BOR.)

This aerial photograph, taken on October 17, 1962, illustrates the thin arch construction used for the Flaming Gorge Dam. On the left is the area where the transmission lines will be placed, and on the right is the concrete batch plant. The concrete placement has reached the maximum height of 502 feet on a portion of the dam. (Courtesy of BOR.)

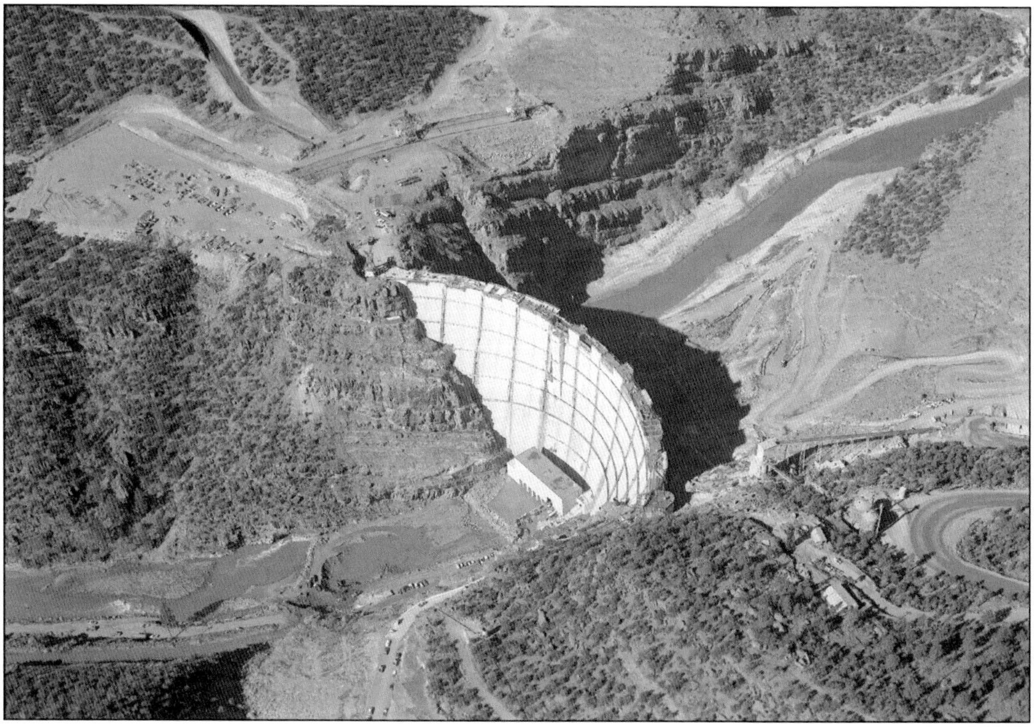

On November 15, 1962, Jean R. Walton (left), project construction engineer, and Douglas D. Baker, project manager for Arch Dam Constructors, shake hands as they prepare to pull the cord to empty the last bucket of concrete. After the placement of the last bucket, the dam and powerhouse will contain approximately one million cubic yards of concrete. (Courtesy of BOR.)

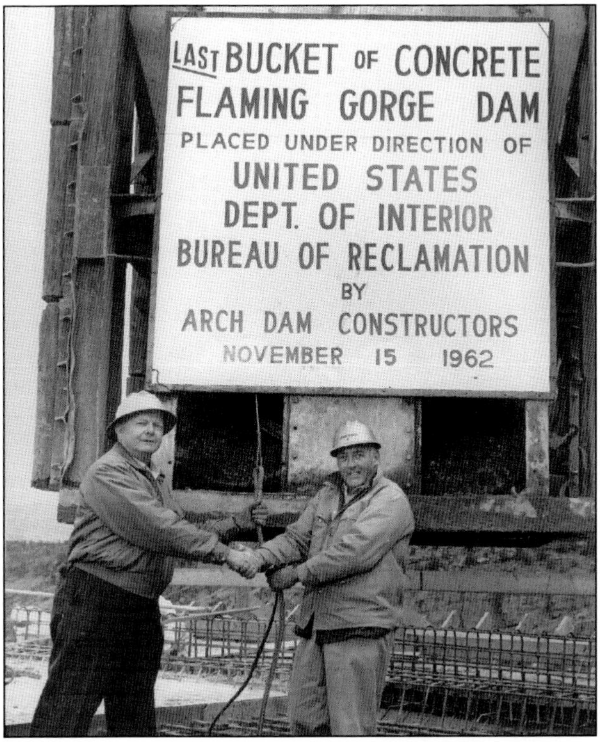

This view is looking upstream and toward the right abutment of the Flaming Gorge Dam. The large bucket can be seen hanging over the rim of the dam on the far right side where the last bucket of concrete was placed. The completed dam has a height of 502 feet above the lowest point in the foundation and a crest of 1,285 feet. This photograph was taken on November 15, 1962. (Courtesy of BOR.)

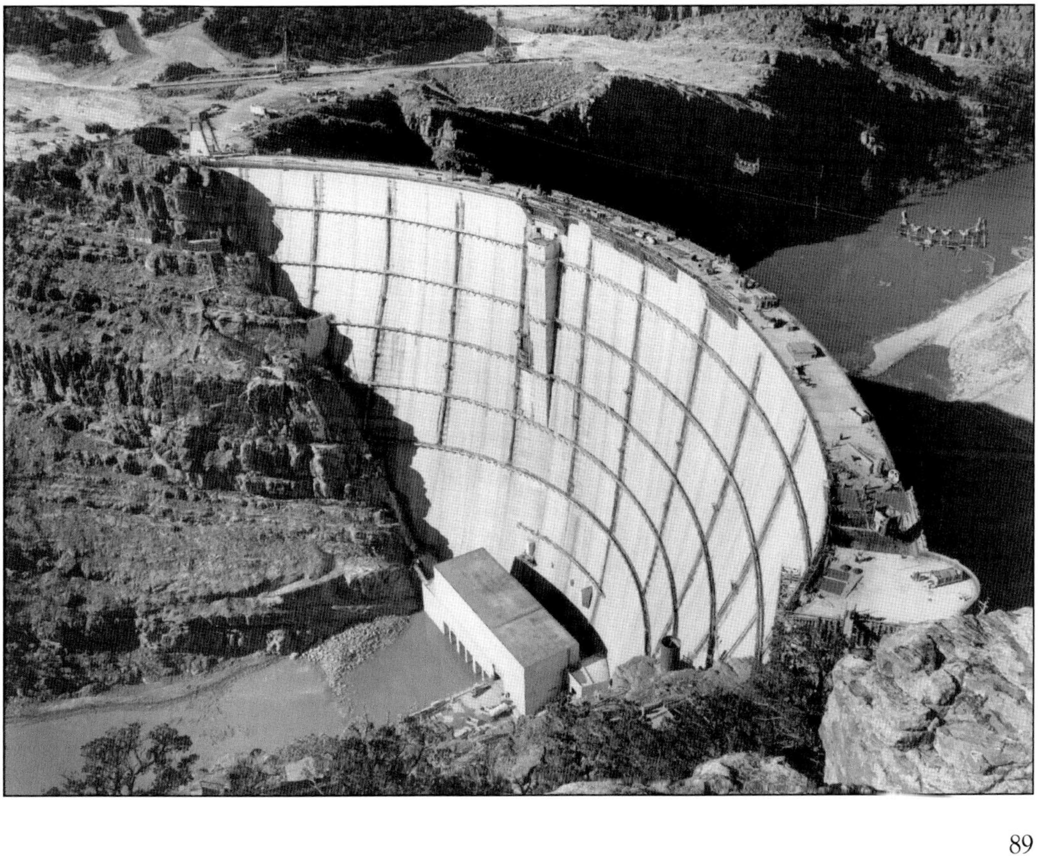

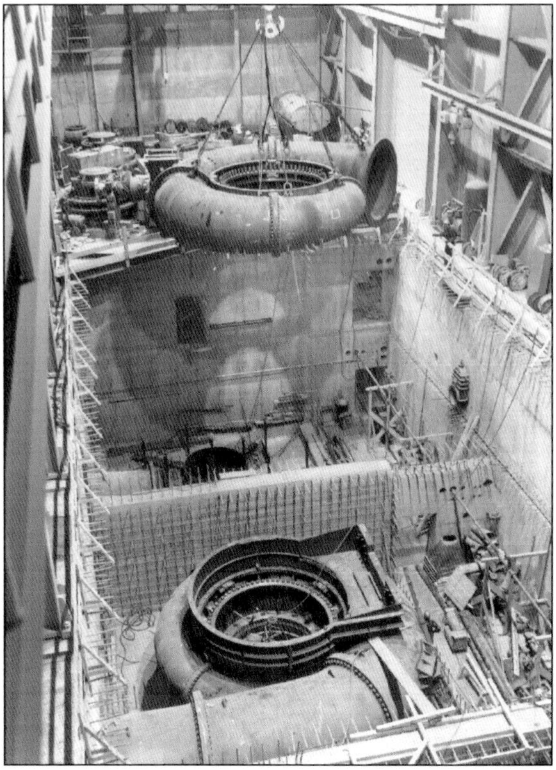

In February 1963, the scroll case is lowered into unit bay No. 3 in the interior of the powerhouse. The scroll case, which has the appearance of a giant snail, directs water to the water wheel, which will be placed in the middle. The eight-foot-diameter case weighs 122,834 pounds. It will be aligned level and then embedded in concrete. (Courtesy of BOR.)

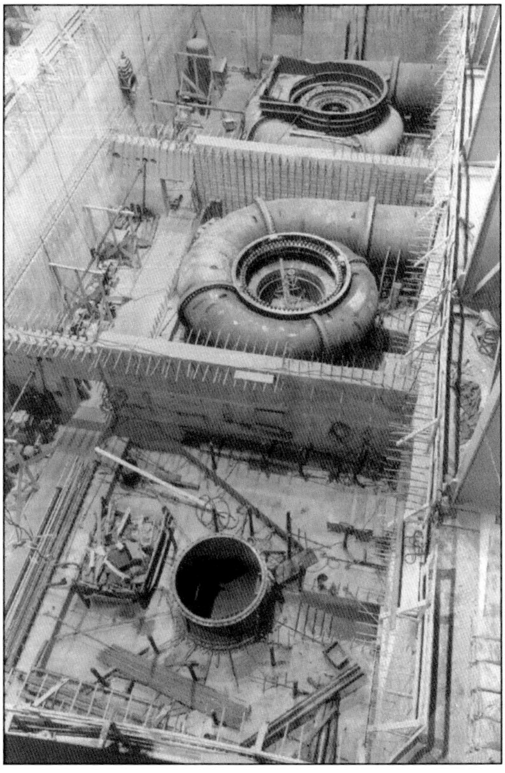

Shown here is the inside of the powerhouse looking into the unit bays. Progress is being made on the installation of the spiral scroll cases in unit bays no. 1 and no. 2 and of the draft tube liner in unit bay no. 3. This photograph was taken on December 31, 1962. (Courtesy of BOR.)

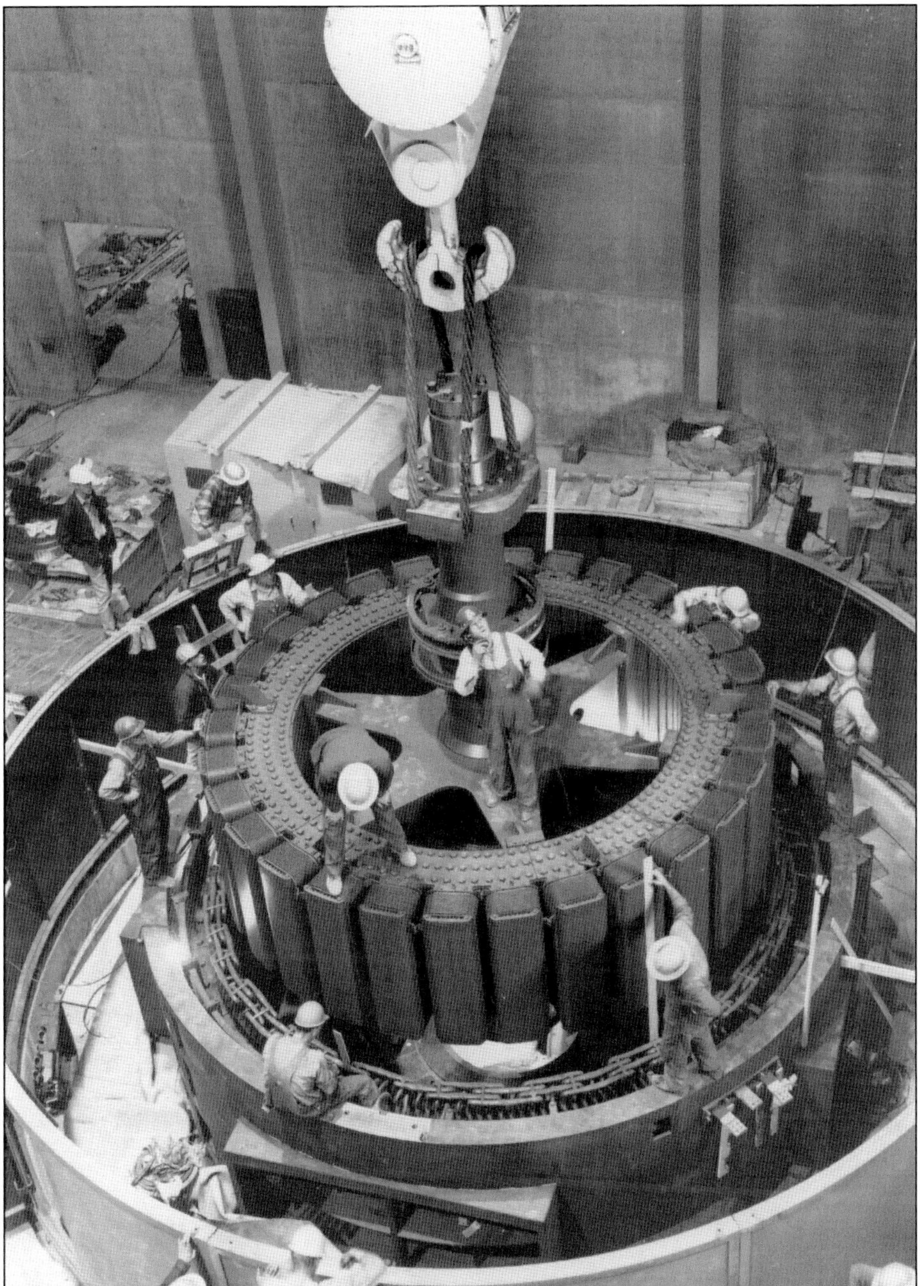

On June 6, 1963, Frank Adams (center), an ironworker from Myton, Utah, calls directions via telephone to the crane operator as crewmen install the revolving rotor for the powerhouse's no. 1 generator. The 300,000-pound overhead crane lowers the 240,000-pound rotor of laminated steel into place. The rotor will rotate at 240 revolutions per minute and generate 36,000 kilowatts of electricity. It will be driven by a 50,000-horsepower turbine operating under a normal water head of 365 feet. This unit was one of three identical units. When all of these are installed, they will provide a total of 108,000 kilowatts of electricity—enough to power 180,000 homes. (Courtesy of BOR.)

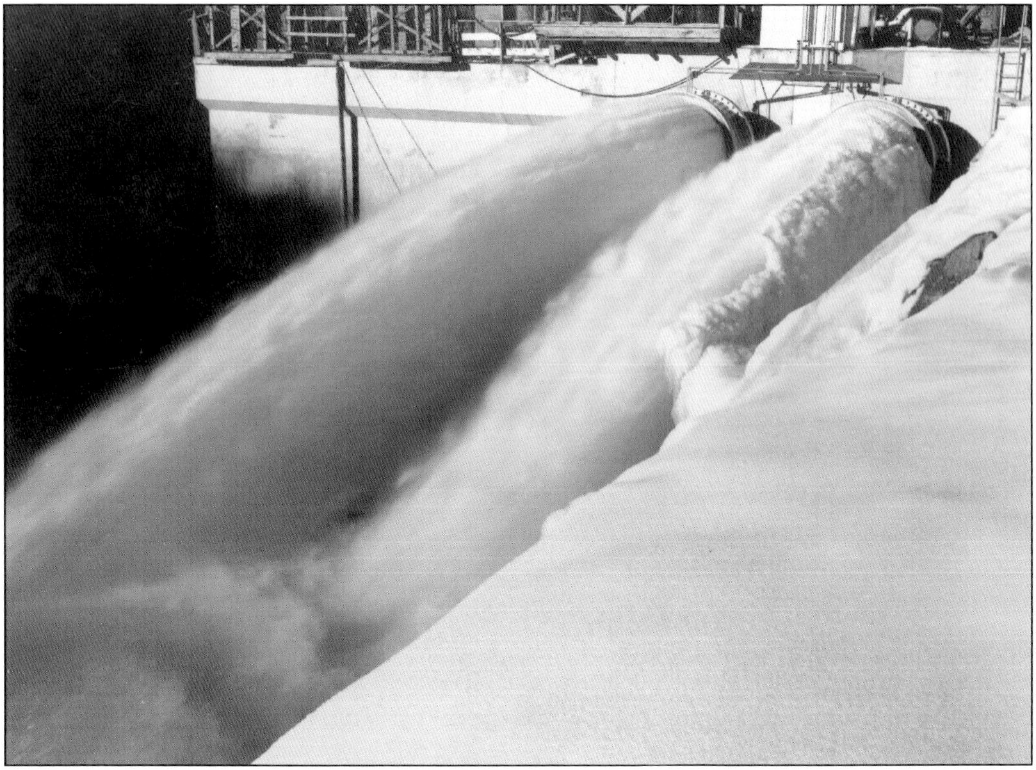

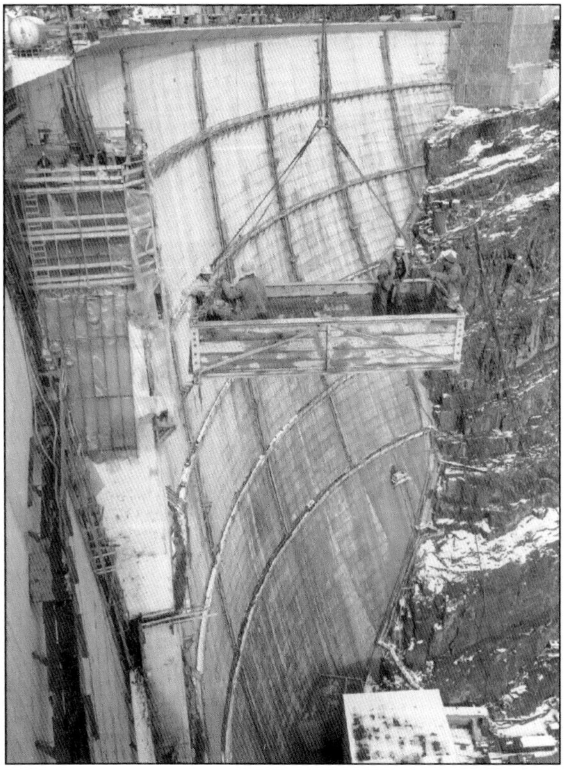

In January 1963, water roars from both jet valves of the river outlet works at Flaming Gorge Dam; an estimated outflow is 360 cubic feet per second. On the right, ice can be seen forming due to subzero temperatures. An official low of 38 degrees below zero was recorded. (Courtesy of UCRHC.)

Surveyors for the Bureau of Reclamation and workmen of the Arch Dam Constructors ride in the skip, inspecting the dam on March 18, 1963. The powerhouse is located about 400 feet below them. The elevator shaft is under construction on the left. In the upper right, the bridge section of the spillway is under construction. (Courtesy of BOR.)

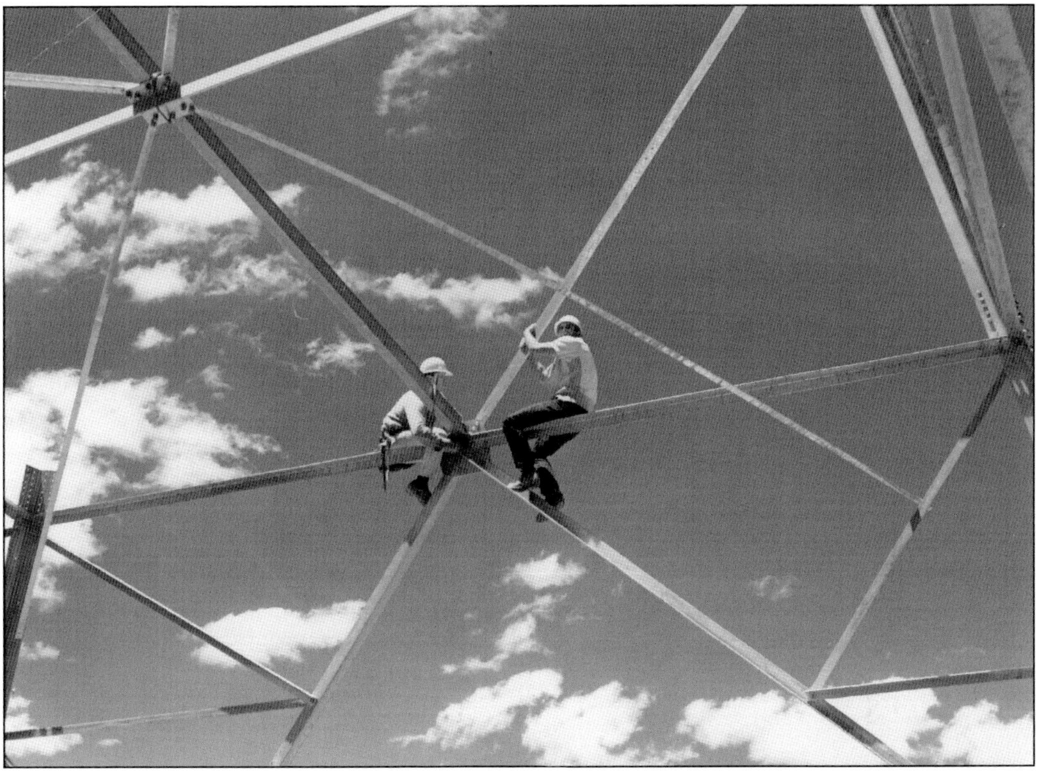

Joe Evans (left) of Modesto, California, and Monk Bailey of Moab, Utah, assemble one of the towers for the switchyard at the Flaming Gorge Dam on June 12, 1963. When completed, the tower will be 141 feet high. (Courtesy of BOR.)

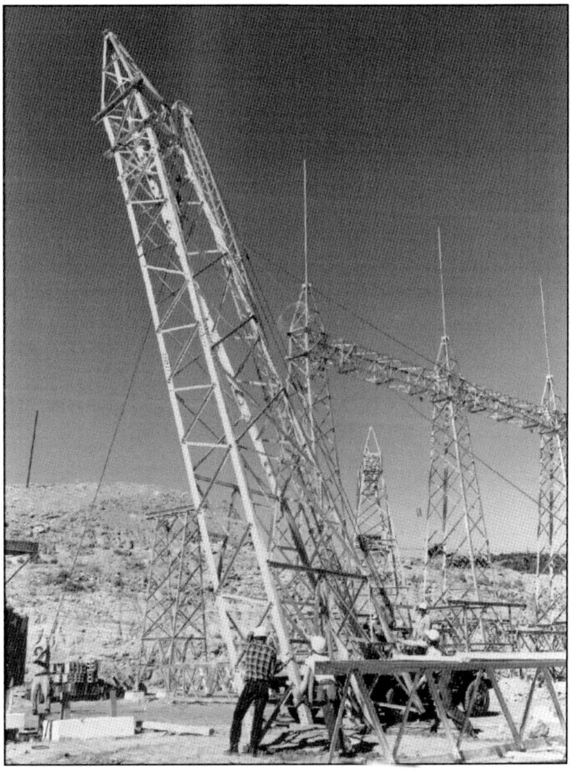

On March 21, 1963, a 53-foot, 138-kilovolt takeoff tower is raised into place at the switchyard at Flaming Gorge Dam. At the time of this photograph, many of the steel structures have been erected for the powerhouse and transmission line. (Courtesy of BOR.)

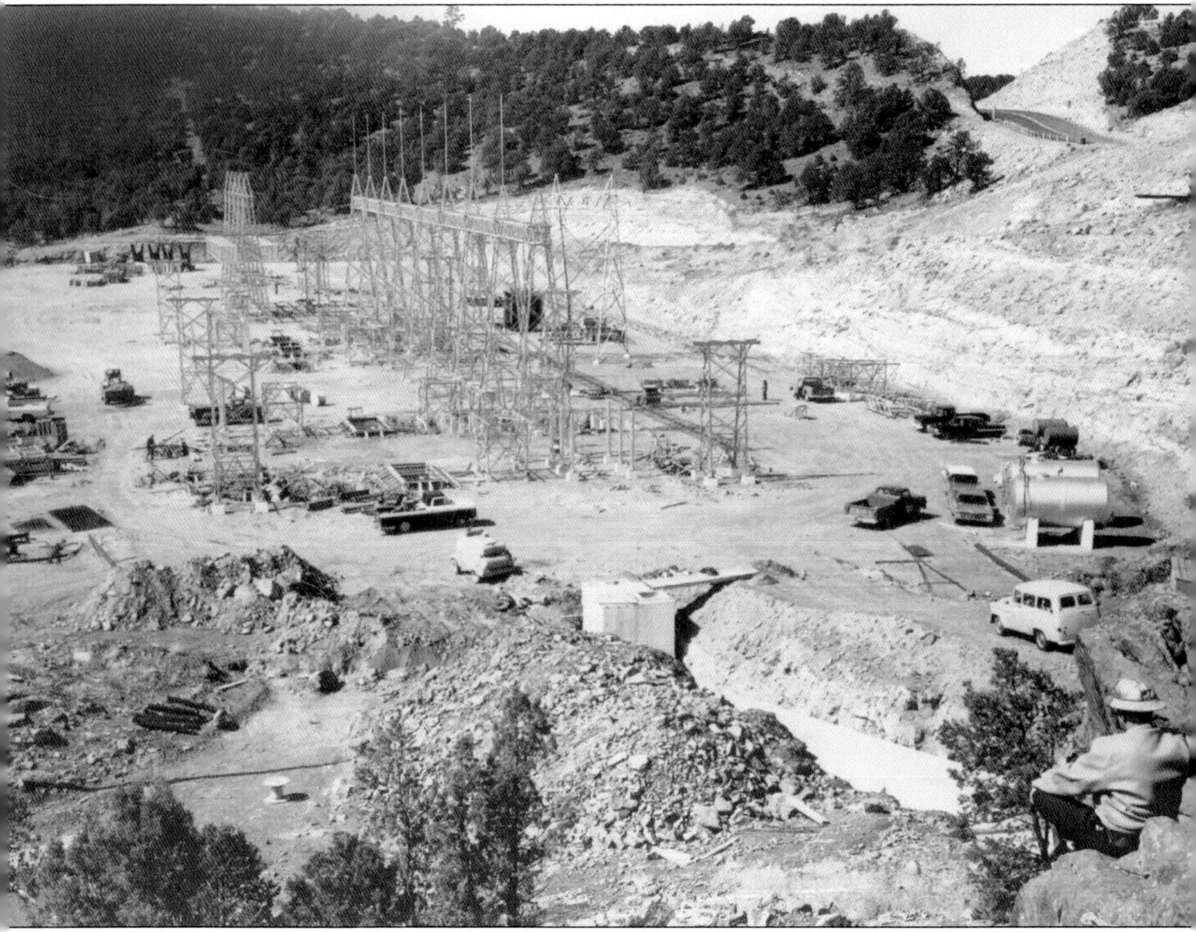

The switch station construction at Flaming Gorge Dam forms a lattice of steel. Shown here is the 138-kilovolt, 69-kilovolt, and 25-kilovolt buss with takeoff structures and associated equipment. In the lower right is the concrete control cable tunnel, which will house the cables leading from the power plant. A large part of the structural steel is already in place. Gunther, Shirley and Lane was contracted to complete the powerhouse and switchyard. The large generator in the powerhouse is scheduled to be turned on in October 1963, at which time power will flow to the switchyard. The lines will be energized to deliver electricity to distant cities. This photograph was taken in April 1963. (Courtesy of BOR.)

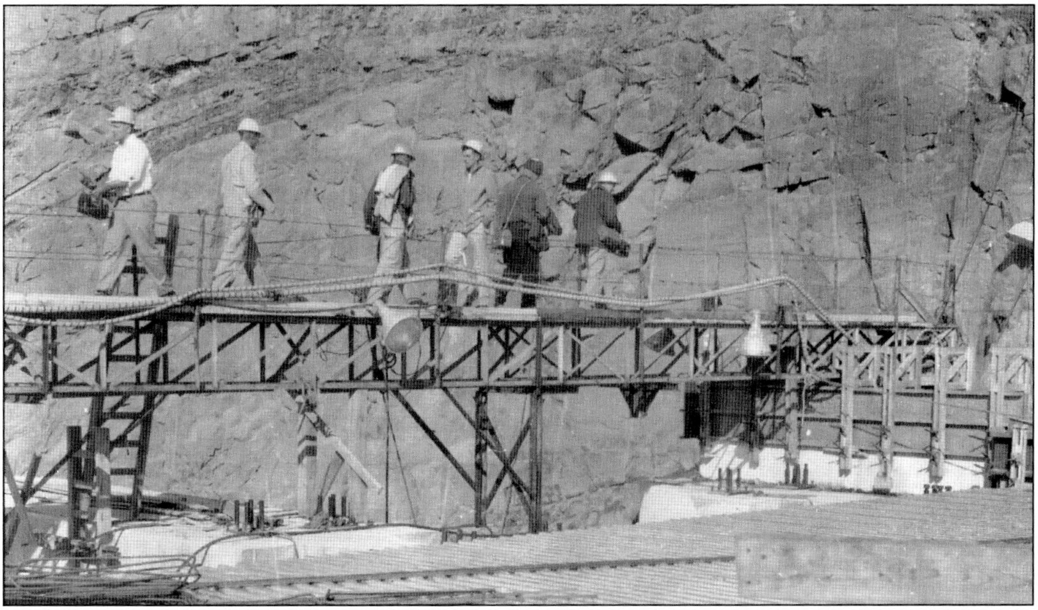

Construction workers can be seen carrying their lunch boxes across the catwalk to their jobs on the Flaming Gorge Dam in May 1963. The catwalk was used during construction of the bridge over the spillway. (Courtesy of UCRHC.)

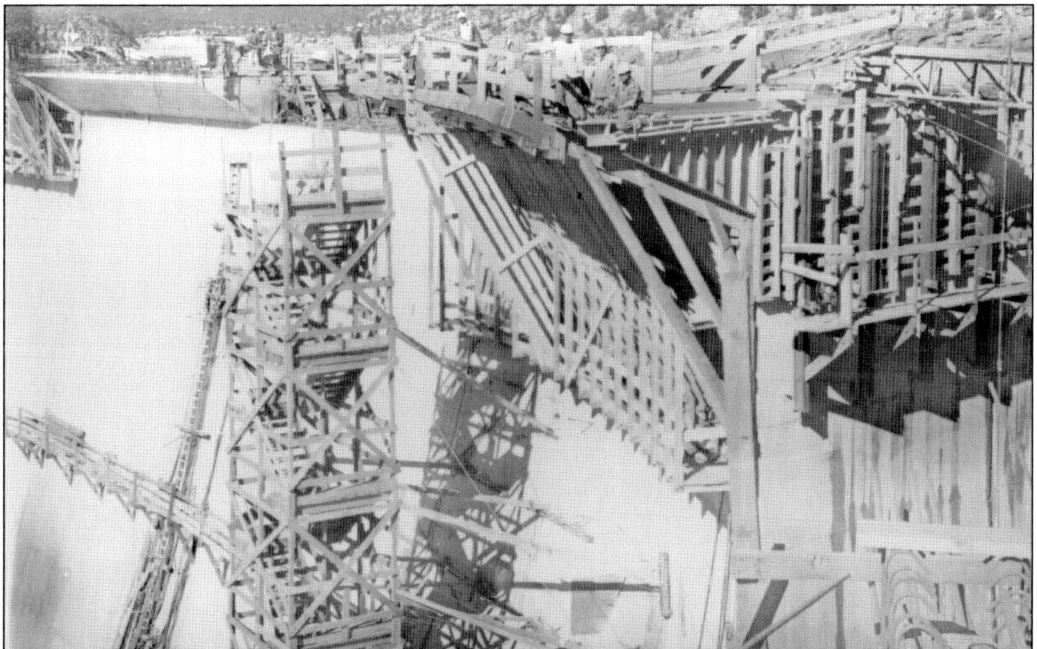

Men are seen at work on top of Flaming Gorge Dam in November 1962. The concrete work is nearly completed, but work continues on the spillway, seen at right. (Courtesy of UCRHC.)

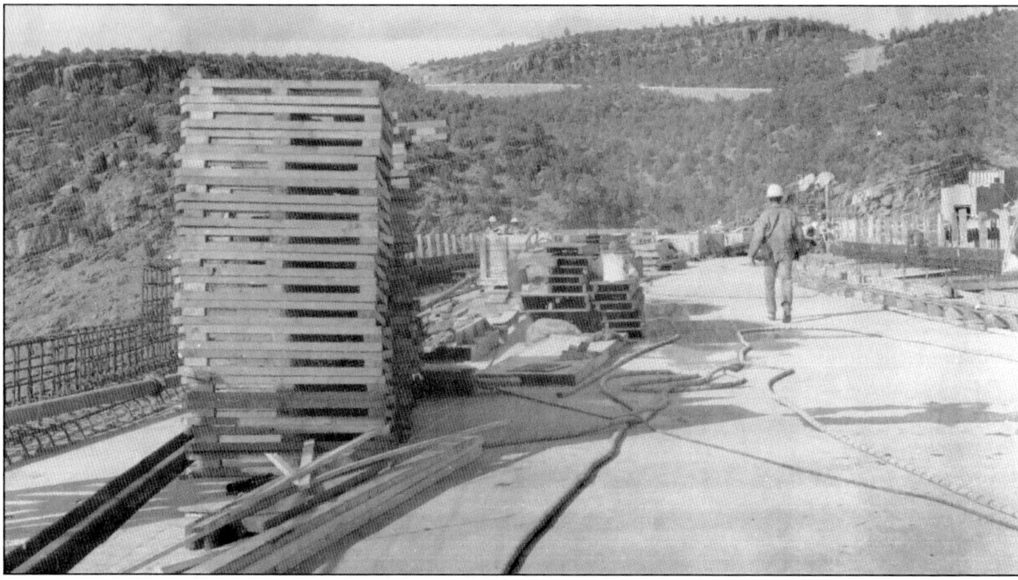

Road construction crew members work on a section of the cement highway across the top of Flaming Gorge Dam in September 1963. When completed, the road will connect to a highway that offers one of the most scenic routes in the area. (Courtesy of UCRHC.)

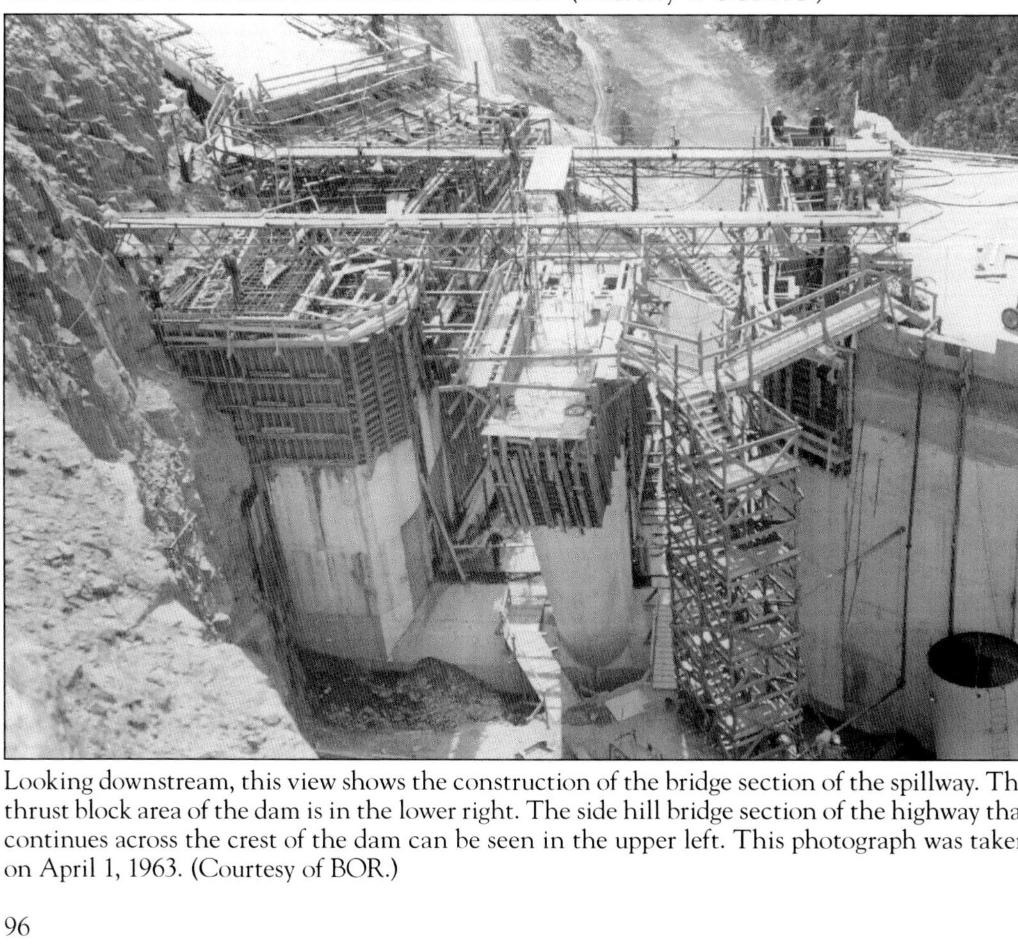

Looking downstream, this view shows the construction of the bridge section of the spillway. The thrust block area of the dam is in the lower right. The side hill bridge section of the highway that continues across the crest of the dam can be seen in the upper left. This photograph was taken on April 1, 1963. (Courtesy of BOR.)

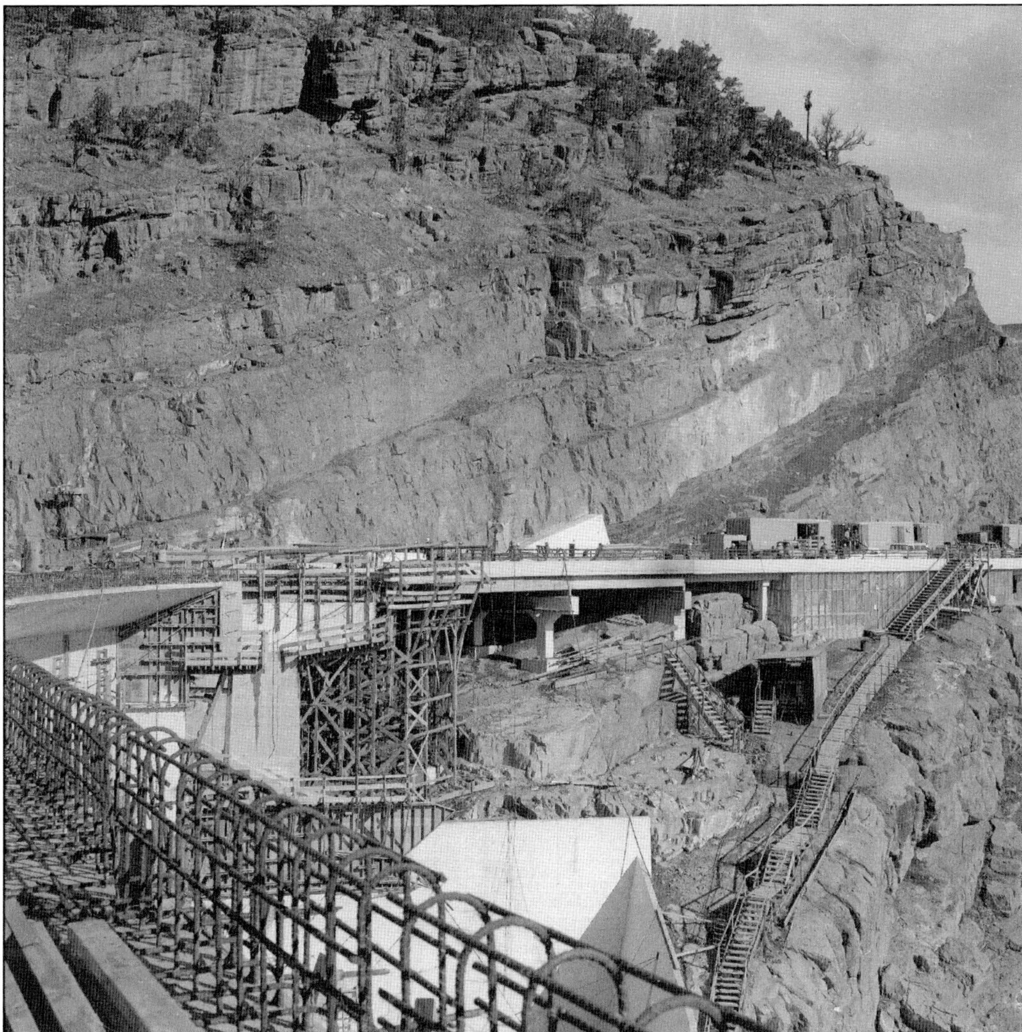

Work continues on the highway that runs over the crest of the Flaming Gorge Dam and on the side hill bridge over the spillway. Scaffolding forms are in place for concrete placement for support of the two-lane highway. Railing is being placed along the edges of the dam for safety. Highway 191 will cross over the dam and wrap around the mountain before continuing on to the community of Dutch John. From there, the highway will give access to recreation areas on the east side of Flaming Gorge Reservoir and provide travelers with a route to Rock Springs, Wyoming. Travel across the dam opened in October 1963. (Courtesy of UCRHC.)

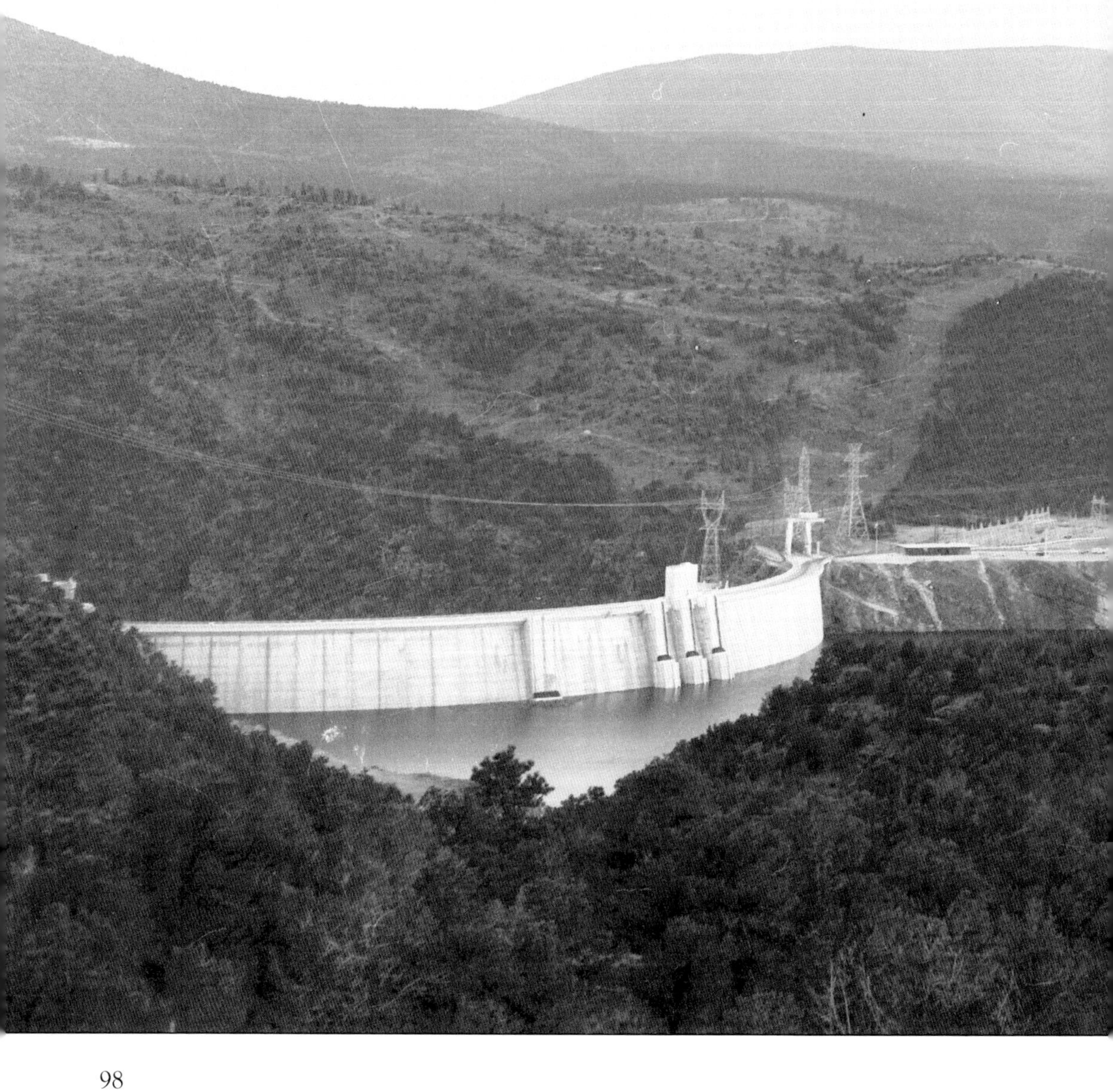

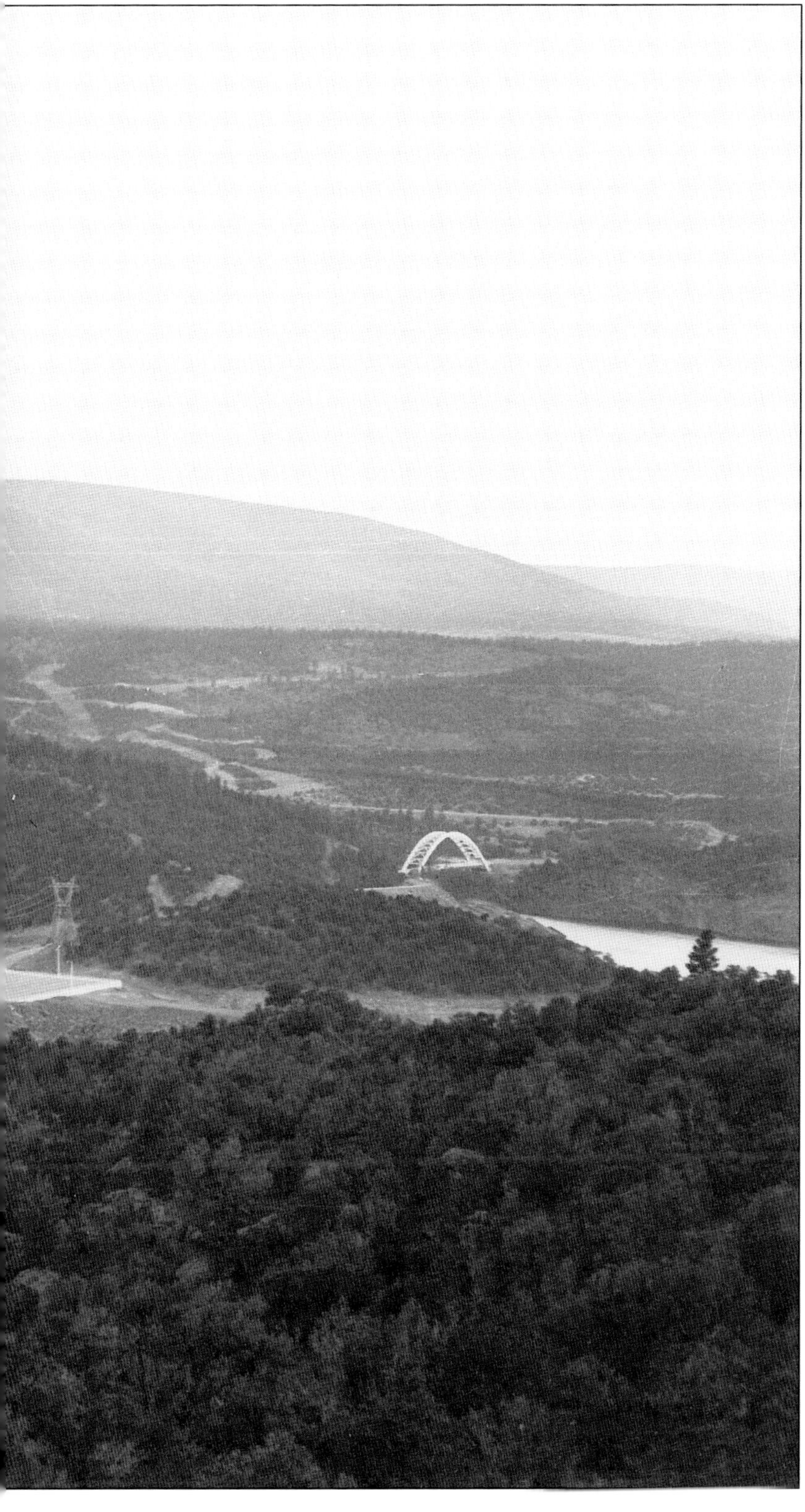

This panoramic view showcases the new Flaming Gorge Dam, the power plant switchyard, and the Cart Creek Bridge. The Green River is beginning to back up against the new dam; eventually, the river would supply a reservoir nearly 100 miles long. The efforts of the Upper Colorado River Storage Project have finally paid off in conserving the waters of the Green River for future use. The Bureau of Reclamation administers the operation and maintenance of the Flaming Gorge Dam, the power plant, and the reservoir. (Courtesy of UCRHC.)

Flaming Gorge Dam — History	
Construction Authorized	April 11, 1956
Award First Construction Contract (Temporary Access Road)	Jan. 4, 1957
Award Prime Contract	June 18, 1958
Diversion of Green River Around Damsite	Aug. 17, 1959
First Bucket of Concrete	Sept. 8, 1960
Last Bucket of Concrete	Nov. 15, 1962
Start of Storage, Flaming Gorge Reservoir	Dec. 10, 1962
First Power Generation	Sept. 27, 1963
Last Generator Completed	Feb. 12, 1964
Dedicated by Mrs. Lyndon B. Johnson	Aug. 17, 1964

Flaming Gorge Dam	
Height Above Bedrock	502 Feet
Height Above Original River Channel	455 Feet
Maximum Thickness	151 Feet
Thickness at Crest	27 Feet
Crest Length — (Arc Length at Axis of Dam)	1,285 Feet
Volume of Concrete	
(Dam Only)	986,644 Cu. Yds.
Powerplant & Misc.	54,843 Cu. Yds.
Total Concrete	1,041,487 Cu. Yds.
Cost of Dam & Reservoir	$49,600,000
Cost of Powerplant & Switchyard	$15,400,000
Cost of Dam, Powerplant, Switchyard, Movable Property, Structures, etc.	$65,300,000
(of which 96% will be repaid U.S. Treasury from sale of power, 76% of total with interest)	

The figures shown here give the final results and dimensions of the Flaming Gorge Dam, the Bureau of Reclamation's large river storage project. (Courtesy of BOR.)

Seven

DEDICATIONS AND DIGNITARIES

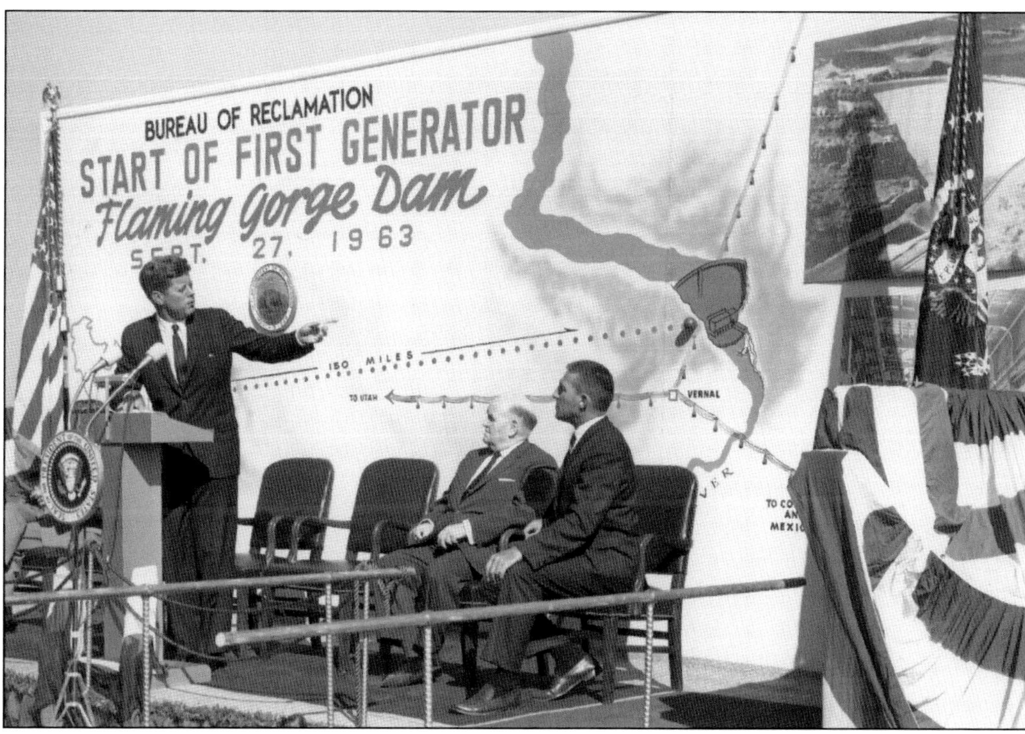

Pres. John F. Kennedy speaks during a brief ceremony in which he pressed the key that started the first generator at Flaming Gorge Dam. Shown here are, from left to right, President Kennedy, assistant reclamation commissioner William I. Palmer, and Secretary of the Interior Stewart L. Udall. (Courtesy of BOR.)

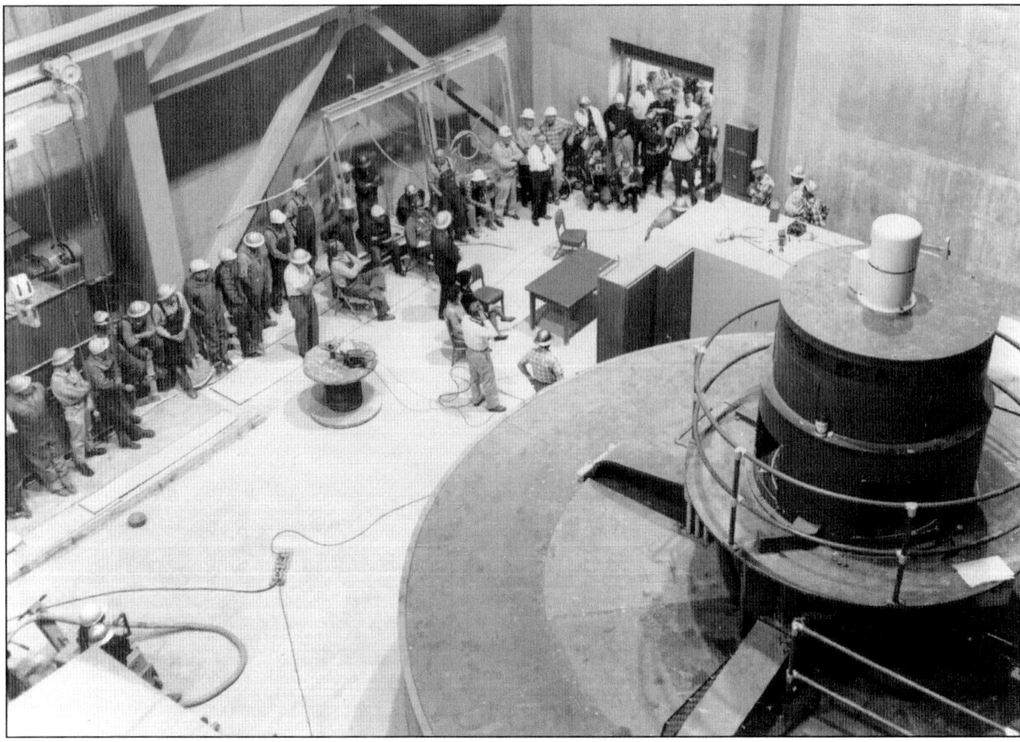

A small group of newsmen gathers with other reporters and project workers in the Flaming Gorge Dam powerhouse as they wait for President Kennedy's signal from the Salt Lake airport to start the generator. All other equipment was stilled to avoid interference with the broadcast. This is the first of three generators that will produce power created by the massive project. (Courtesy of BOR.)

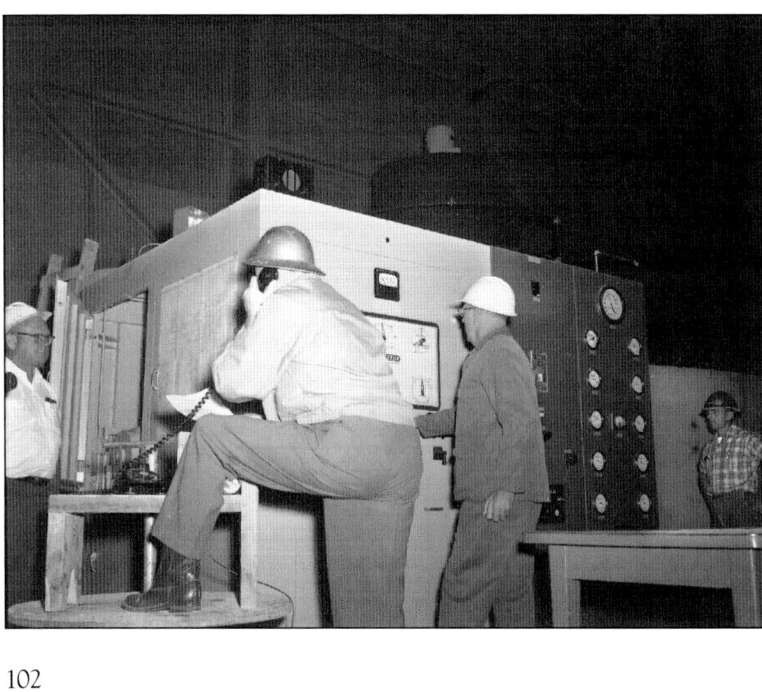

Jean R. Walton (left), Bureau of Reclamation project construction engineer for Flaming Gorge Dam, talks on the telephone to Senator Moss at the Salt Lake airport while waiting for the signal to turn on the generator. Maynard Bulthaus, erection engineer for the firm that made the governor (measuring device), waits to throw the switch on September 27, 1963. (Courtesy UCRHC.)

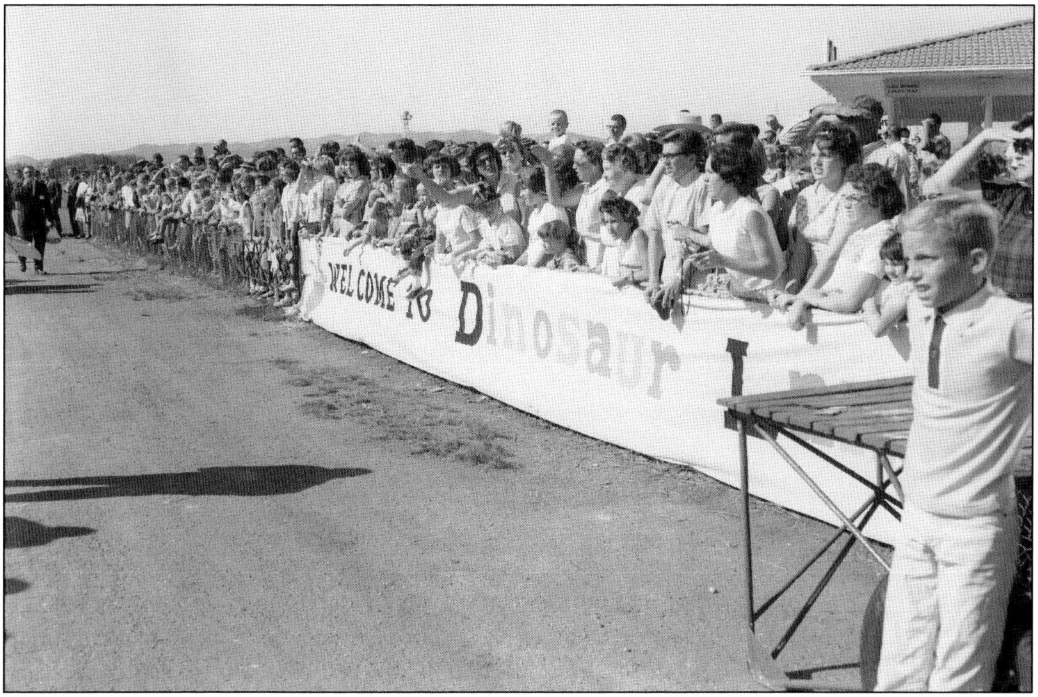

Many spectators are lined up at the Vernal City Airport to await the arrival of First Lady, Lady Bird Johnson. Many came to welcome her to the area for the dedication of the new Flaming Gorge Dam. Spectators stand behind a sign that reads, "Welcome to Dinosaur Land." (Courtesy of UCRHC.)

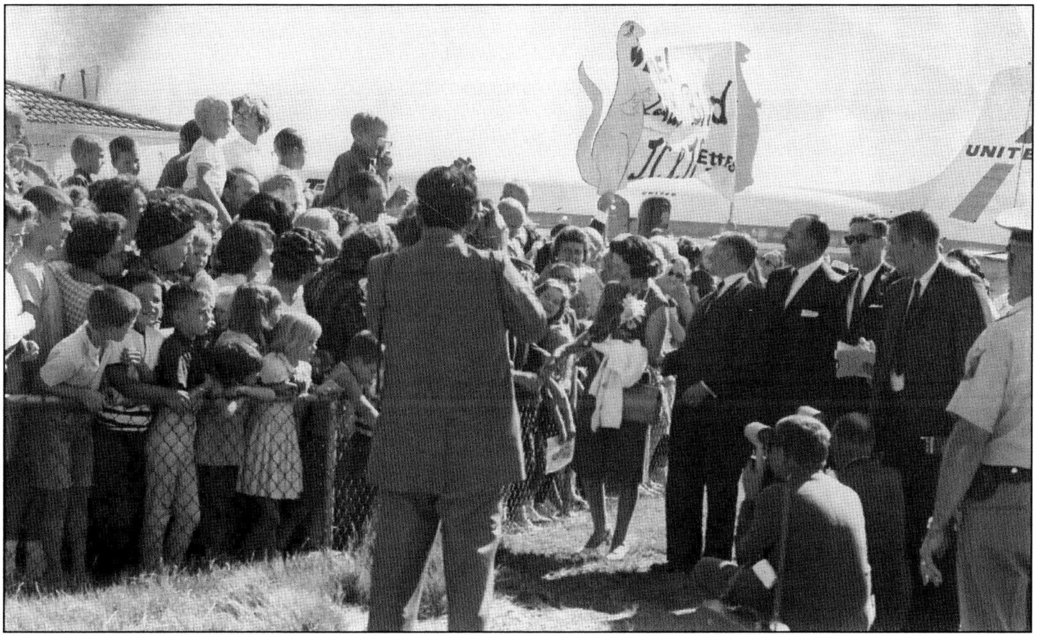

Lady Bird Johnson charms a thousand spectators at the Vernal City Airport. Federal and state officials, reporters, and cameramen surrounded the first lady. After official greetings, she walked along the fence that separates the runway from the lawn and shook hands with everyone she could reach, particularly little children. (Courtesy of UCRHC.)

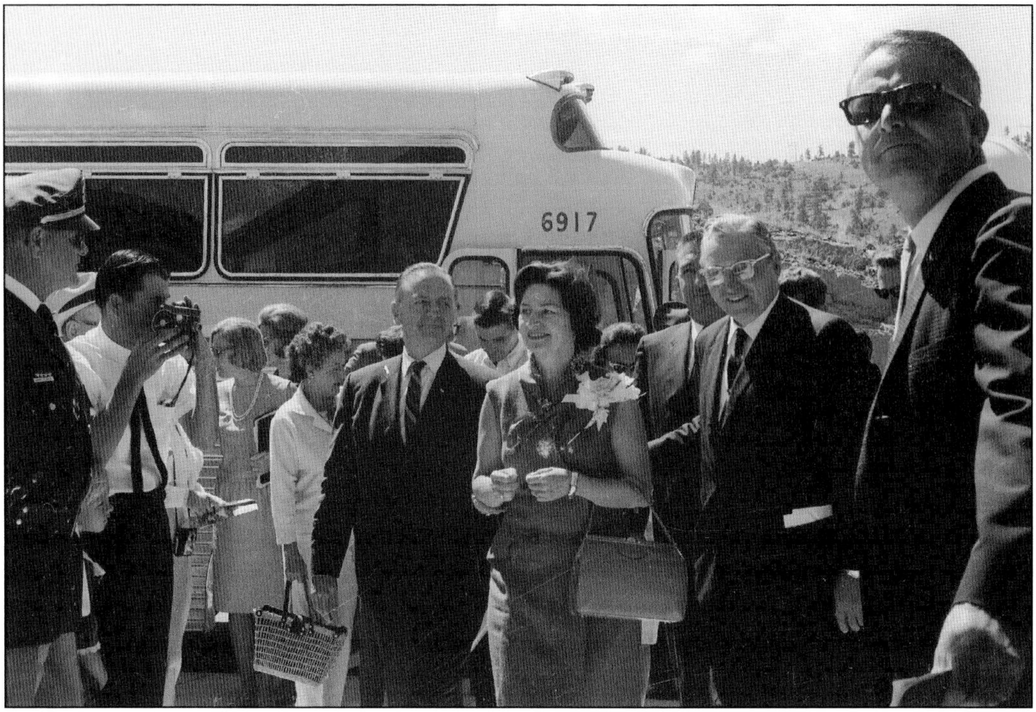

Lady Bird Johnson leaves the bus that transported her to the dedication ceremony of the Flaming Gorge Dam. Many dignitaries are assisting in her arrival. (Courtesy of UCRHC.)

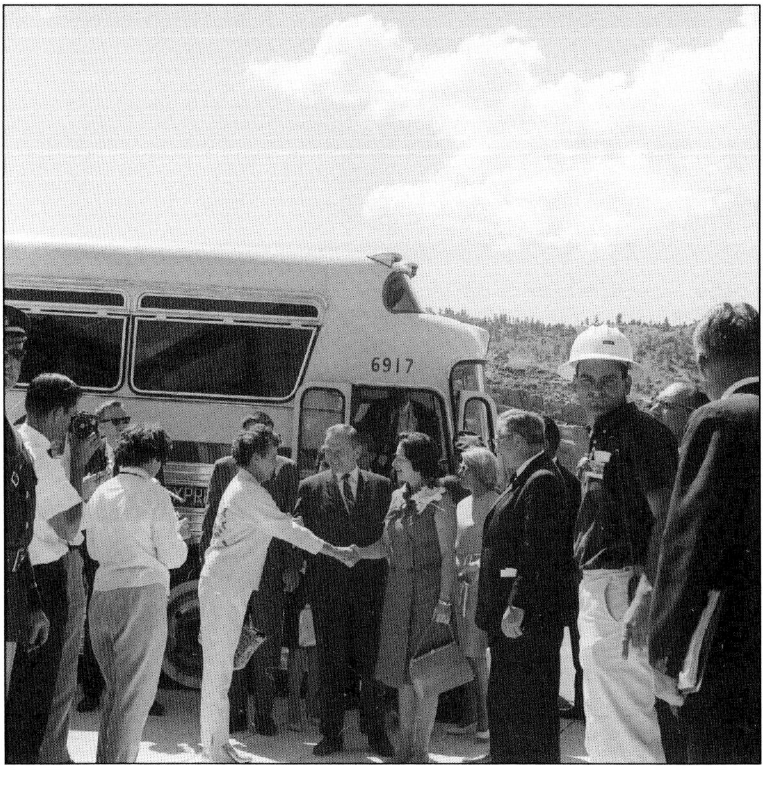

From the moment she stepped from the bus at the Flaming Gorge Dam ceremony, all eyes were on Lady Bird Johnson as she graciously greeted those who were waiting to meet her and hear her speak. Johnson was pinned with a lovely orchid and rose corsage by Ann King prior to her travel at the dedication sight. The chamber of commerce presented her with three plush "Dinahs"; the stuffed dinosaurs represented the Vernal City mascot. (Courtesy of UCRHC.)

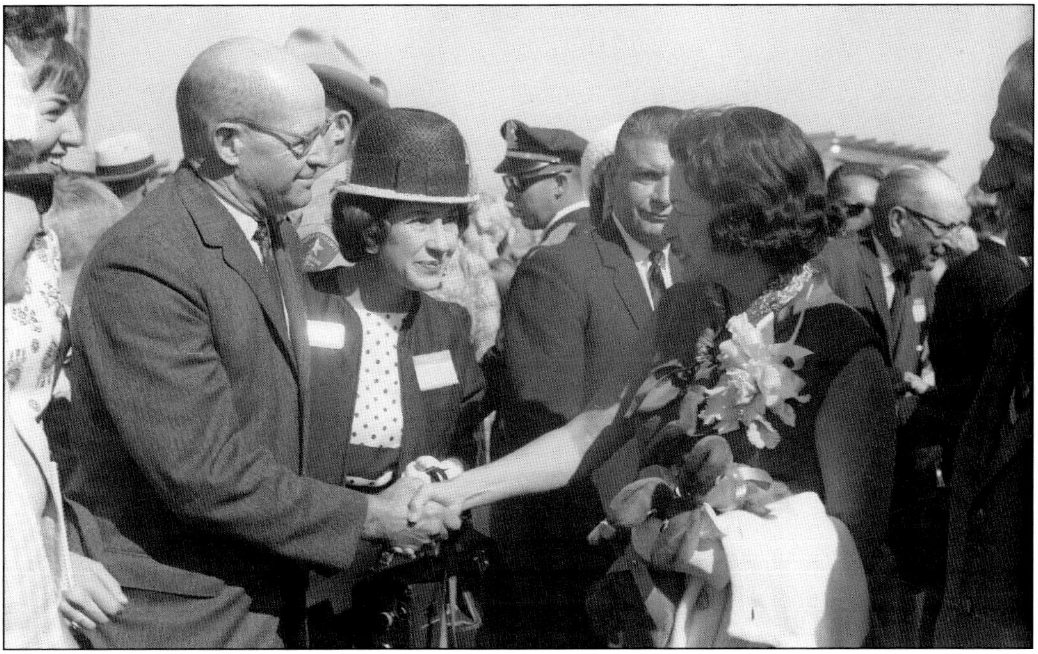

Fred and Theda Washburn (left) and state representative Glenn H. Cooper (center) greet Lady Bird Johnson upon her arrival at the Vernal City Airport. Senator Moss (far right) accompanied her to Vernal. Bry Stringham, former mayor of Vernal and a board member of the Upper Colorado River Storage Project, is behind Johnson (with glasses). (Courtesy of UCRHC.)

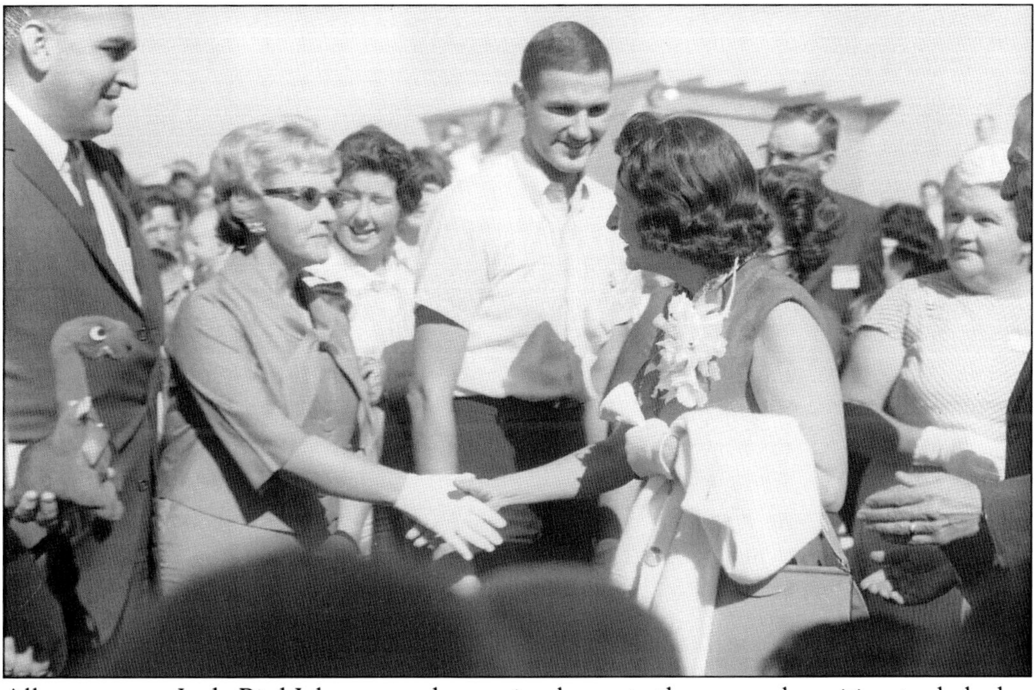

All eyes are on Lady Bird Johnson as she graciously greets those eagerly waiting to shake her hand. She makes her way through the crowd to the dedication ceremony of the Flaming Gorge Dam. (Courtesy of UCRHC.)

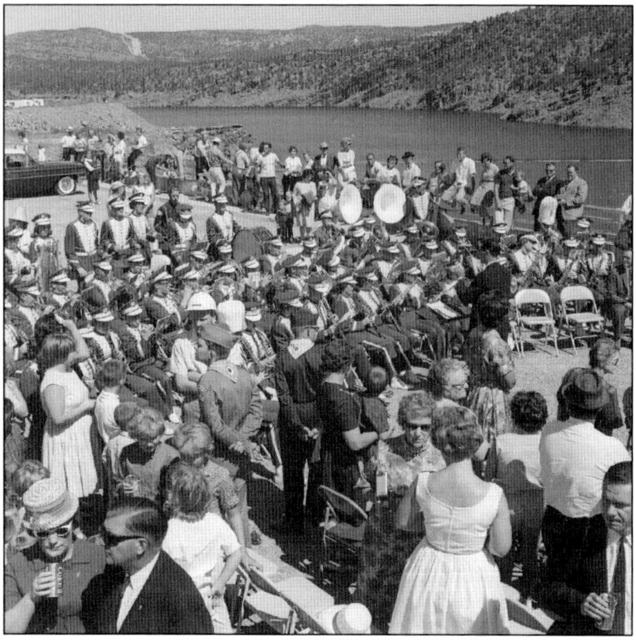

The Uintah High School band provides the music for the Flaming Gorge Dam dedication. Its participation brought a great deal of honor both to the school and the community. Many other spectators are present, eagerly waiting for the ceremonies to begin. (Courtesy of UCRHC.)

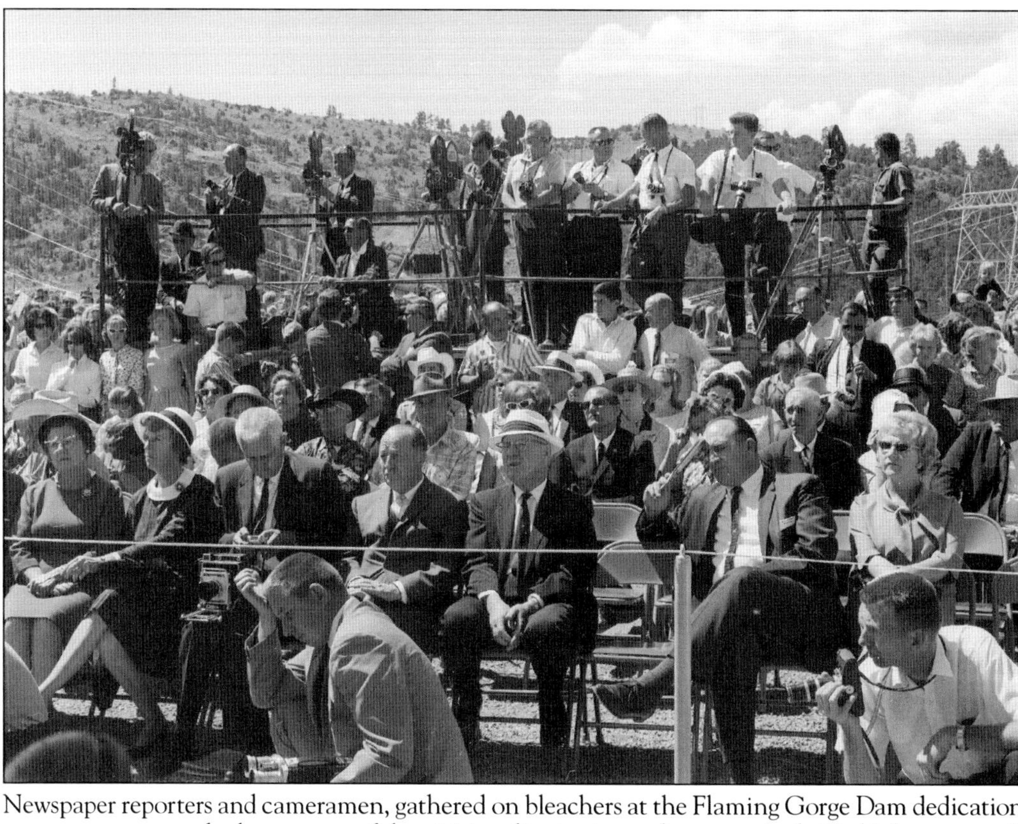

Newspaper reporters and cameramen, gathered on bleachers at the Flaming Gorge Dam dedication, prepare to capture the latest news of the visiting dignitaries at the event. Lady Bird Johnson would soon unveil the plaque and deliver her speech to all who came to support and witness this great event. (Courtesy of UCRHC.)

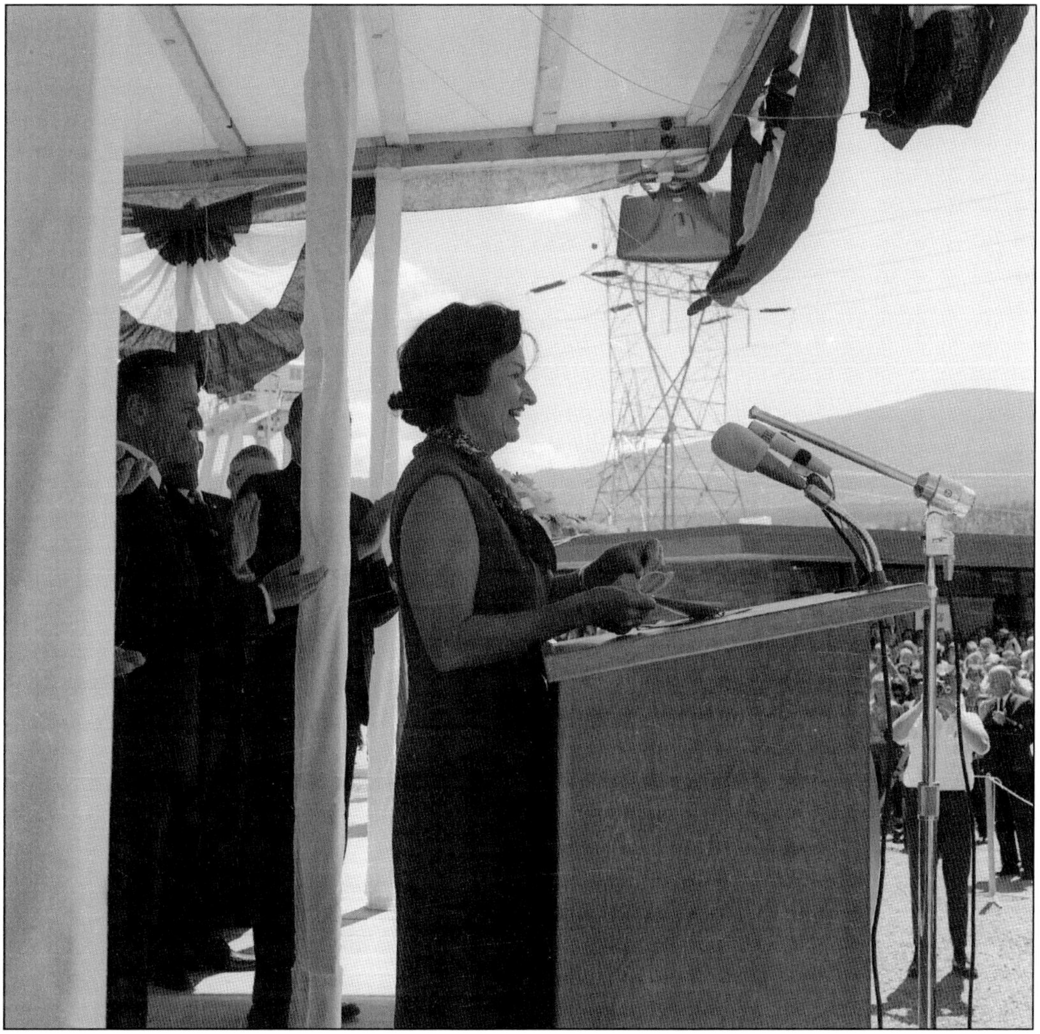

Lady Bird Johnson, standing on a specially constructed platform on top of the 502-foot dam, dedicated the $65 million project. She was the first woman to dedicate one of the West's huge reclamation projects. During her speech, she talked of the spirit of the West and how the early pioneers dreamed big and then worked to make their dreams come true. She referred to John Wesley Powell's vision of a string of dams that would convey water to the desert. "No one can follow the trail I have followed in the last four days without catching the spirit of the West," she said. She talked about the Wilderness Bill and how it would safeguard the beauty of the West. "Let us remember these pioneers. Americans have always felt that the tomorrow of our children should be better than the yesterday of our parents. Let's share a faith in life that the best is yet to come, that we must build our future, not belittle it. In this spirit, and with this wish, I dedicate your Flaming Gorge Dam," she concluded. (Courtesy of UCRHC.)

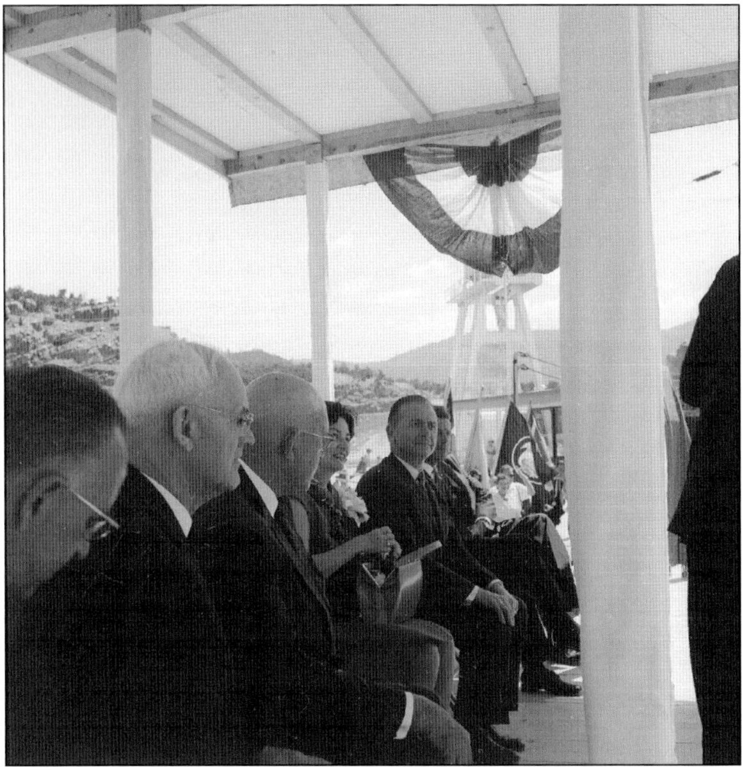

Lady Bird Johnson is seated under the pavilion with other dignitaries. Pictured from left to right are chief forester Edward P. Cliff of the U.S. Forest Service, an unidentified man, Gov. George D. Clyde (vice chairman of the Upper Colorado River Commission), Lady Bird Johnson, Utah senator Frank E. Moss, and Wyoming senator Gale W. McGee. They are waiting to participate in the ceremony to dedicate the Flaming Gorge Dam. (Courtesy of UCRHC.)

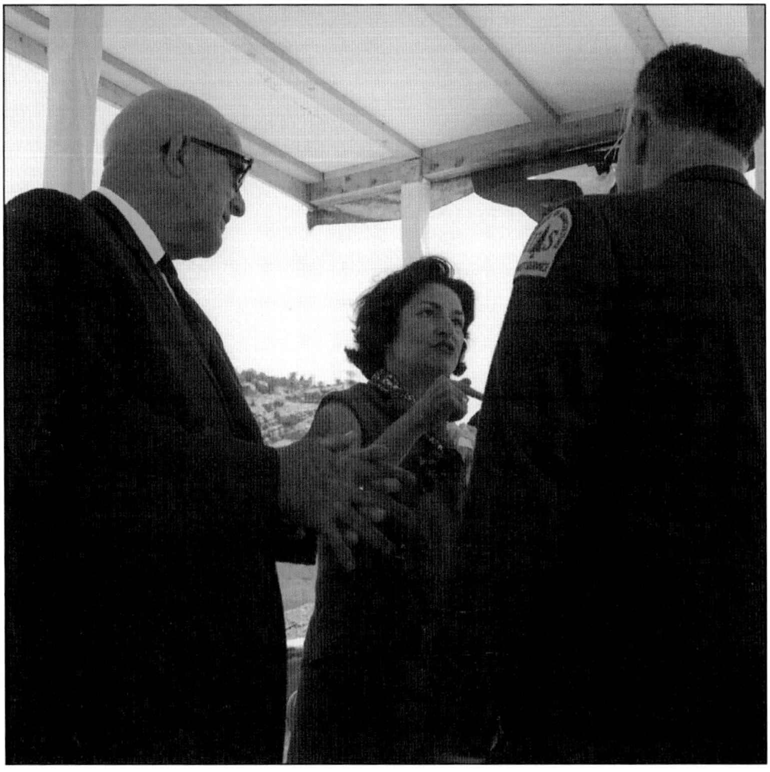

Lady Bird Johnson stands under the pavilion as she talks with a National Park Service ranger and another dignitary. (Courtesy of UCRHC.)

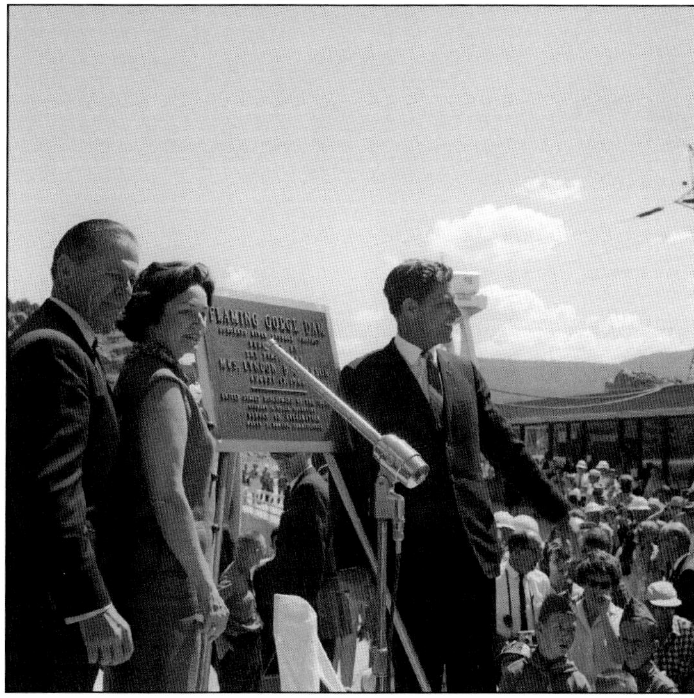

Lady Bird Johnson, Sen. Frank E. Moss of Utah (left), and Sen. Gale W. McGee of Wyoming unveil the plaque at the dedication ceremony for the Flaming Gorge Dam. (Courtesy of UCRHC.)

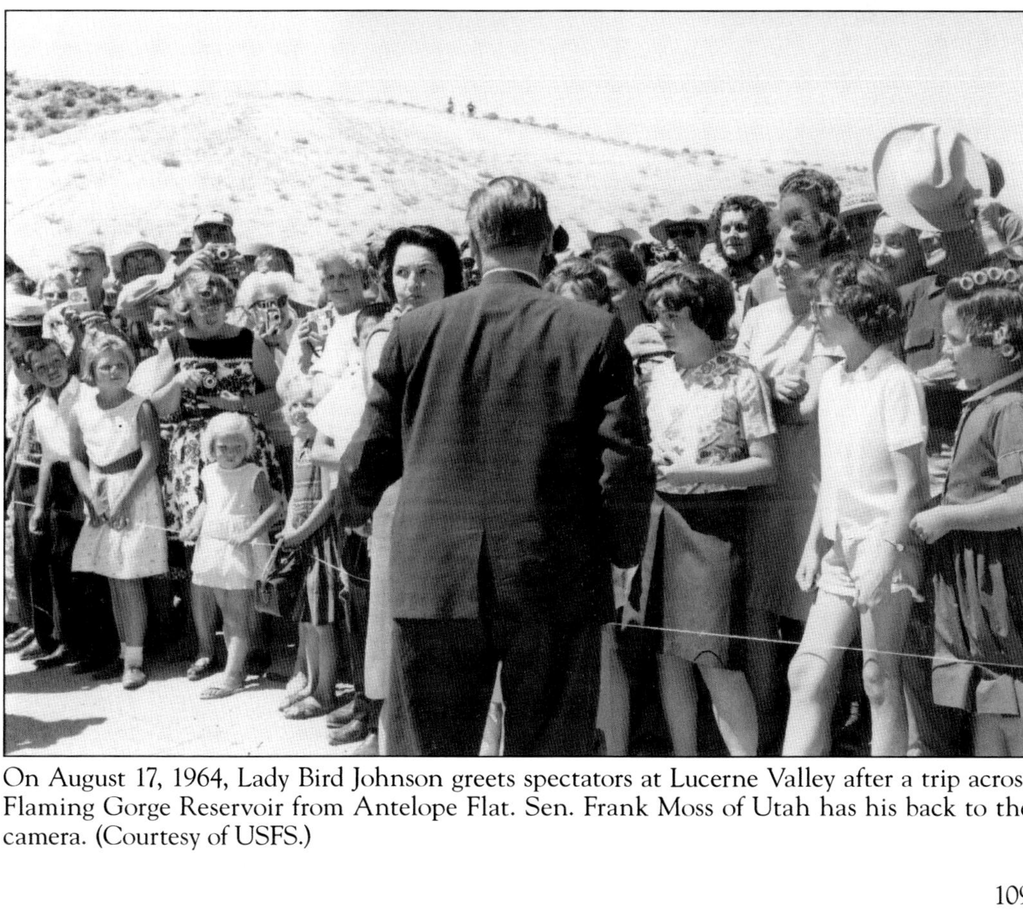

On August 17, 1964, Lady Bird Johnson greets spectators at Lucerne Valley after a trip across Flaming Gorge Reservoir from Antelope Flat. Sen. Frank Moss of Utah has his back to the camera. (Courtesy of USFS.)

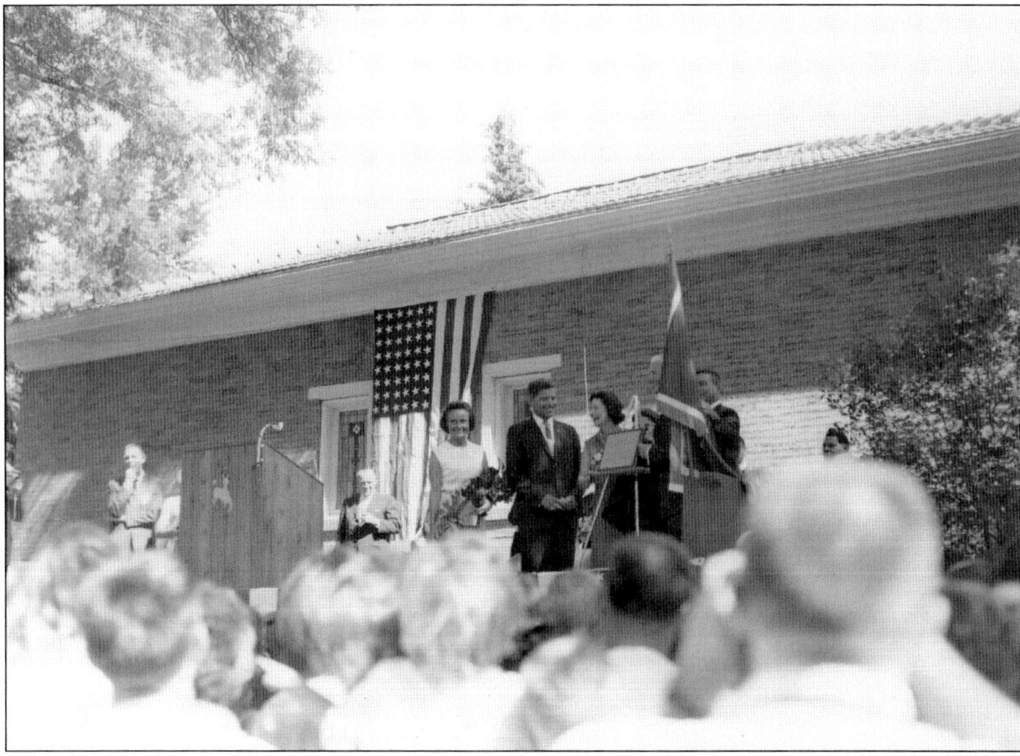

A Flaming Gorge dedication ceremony was held in Green River, Wyoming. Shown here in the foreground are, from left to right, George B. Hartzon, director of the National Park Service; Mrs. and Sen. Gale McGee of Wyoming; Lady Bird Johnson; and Stewart Udall, secretary of the interior (to the left of the flag). The ceremony was held at the Green River Courthouse. (Courtesy of USFS.)

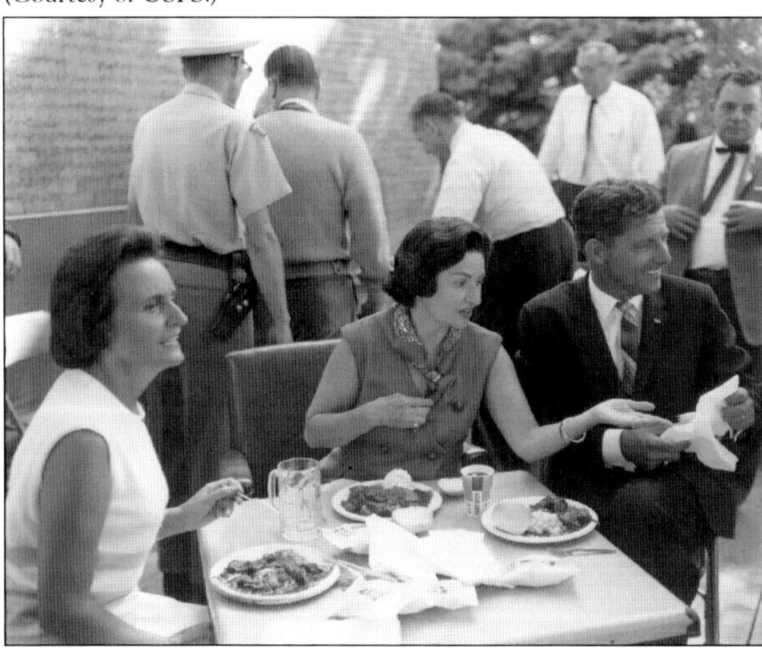

Lady Bird Johnson attends a buffalo barbecue following the dedication ceremonies of Flaming Gorge Dam. Seated at the table with Johnson are Sen. and Mrs. Gale McGee of Wyoming. (Courtesy of USFS.)

Eight

A Sportman's Playground

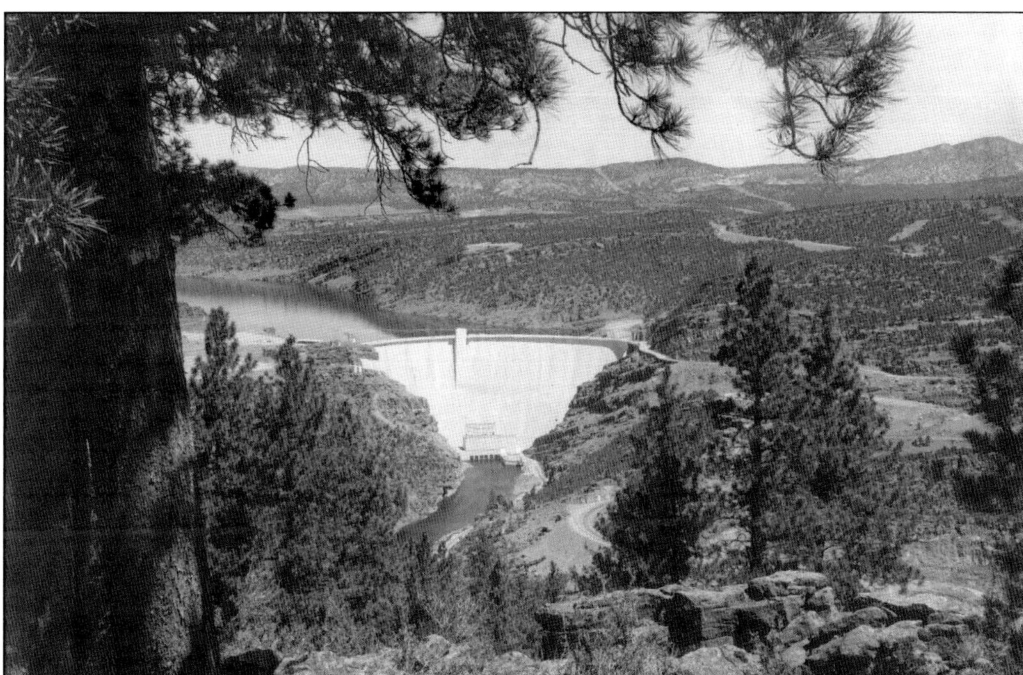

Flaming Gorge Dam brought recreational opportunities for the outdoor enthusiast, and the region became a sportsman's playground. The reservoir provided many picturesque bays and inlets that attracted boaters, water-skiers, and fishermen. The Green River below the dam later became a well-known "blue ribbon" fishery and offered rafting, canoeing, kayaking, and sightseeing. Campgrounds, boat ramps, and picnic areas were also developed in the area. (Courtesy of BOR.)

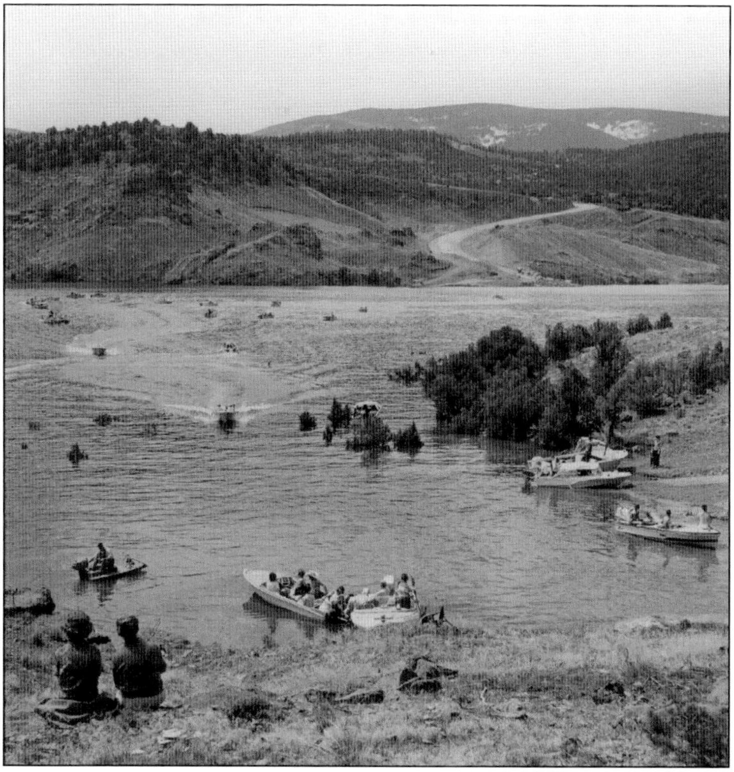

Many boaters flocked to the newly developed Flaming Gorge Reservoir. On this particular day, 60 to 70 boats of the Southwestern Wyoming Boating Club can be seen. Wyoming fishermen and boaters looked forward to the project being completed. Previously, a day of boating or fishing meant a drive of 100 miles or more before their crafts could even be launched. (Courtesy of BOR.)

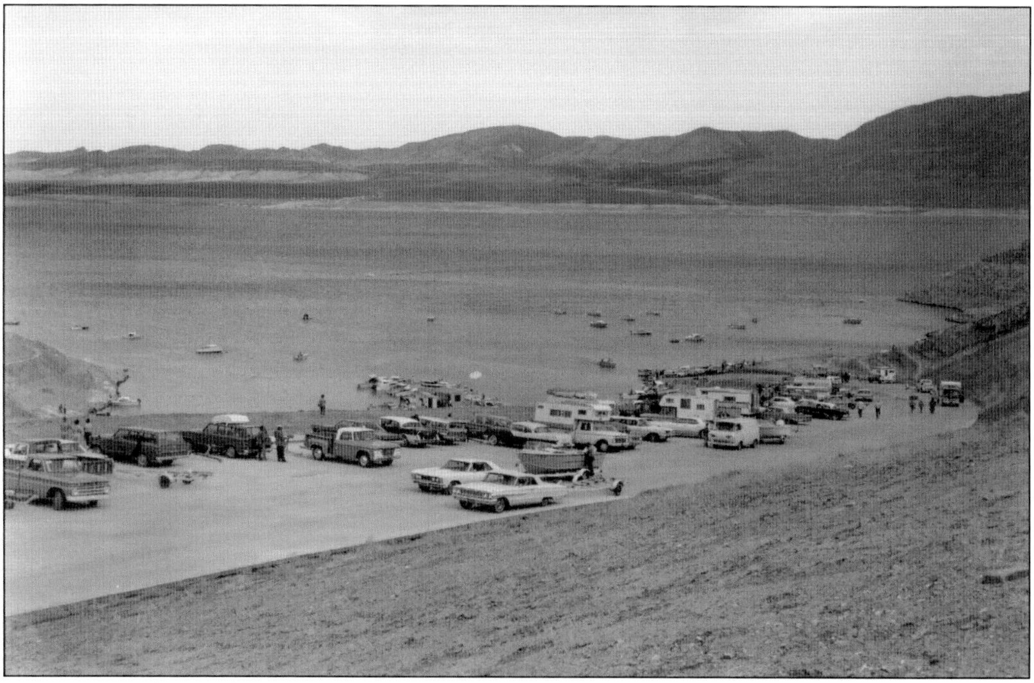

Visitors park their vehicles along the boat ramp at the Lucerne Marina near Manila, Utah. As is evident here, many outdoor enthusiasts enjoyed fishing, boating, and swimming on the newly developed reservoir. (Courtesy of USFS.)

Fishing on the Green River is a favorite pastime for visitors to the Flaming Gorge Reservoir. At the lower end of the diversion tunnel, Brent Feltch of Vernal lands a large squawfish. During the spawn, these fish congregated in large numbers in this fishing hole, much to the delight of fishermen. (Courtesy of Merlin Sinfield.)

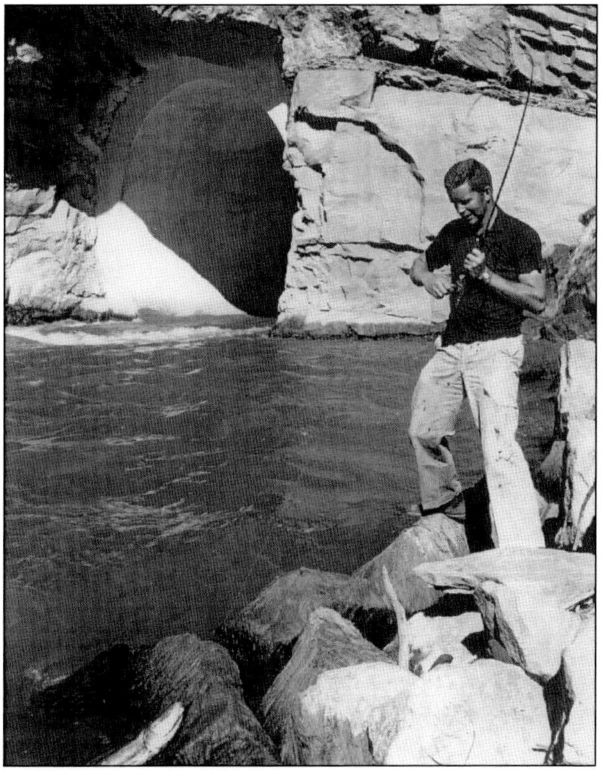

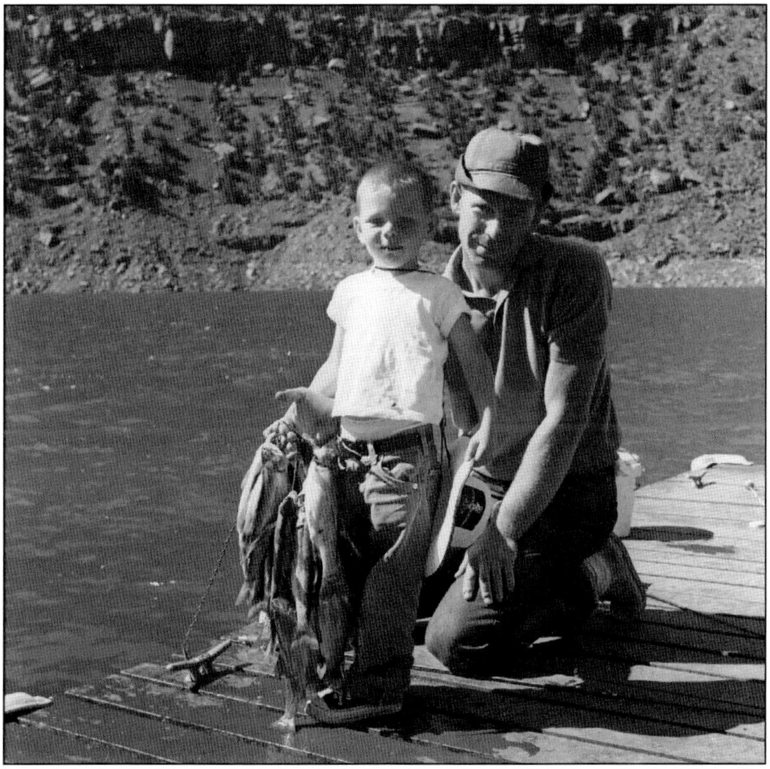

A proud father and son share in a fun day of fishing on the Flaming Gorge Reservoir. Here, they show off their catch of the day. (Courtesy of UCRHC.)

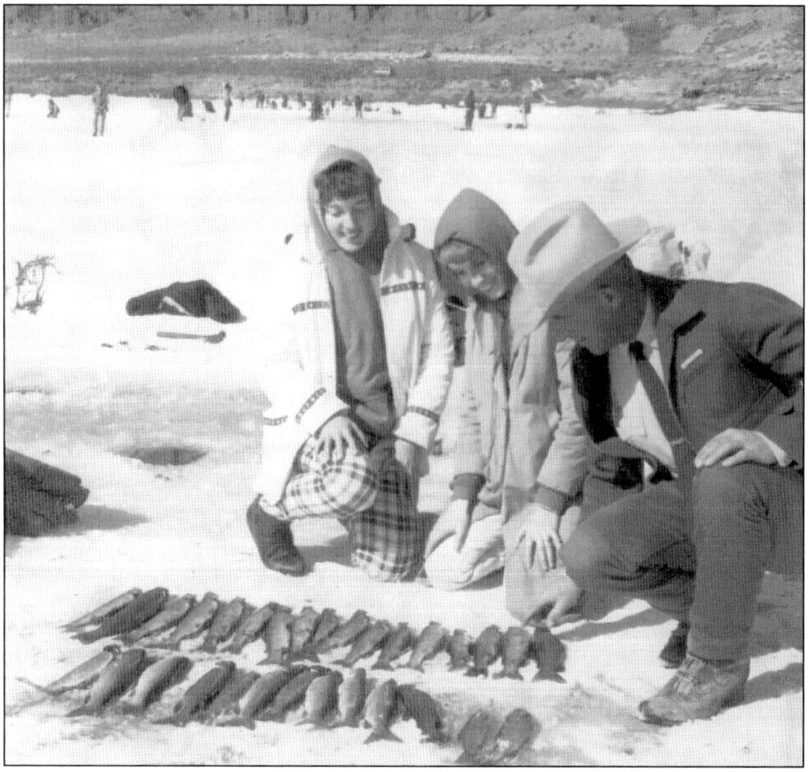

Winter fishing is very popular. Two young ladies from Salt Lake City, Utah, are all smiles as they count their catch from the Flaming Gorge Reservoir. Fish and game officer Dick Bennett admires their day's work. (Courtesy of UCRHC.)

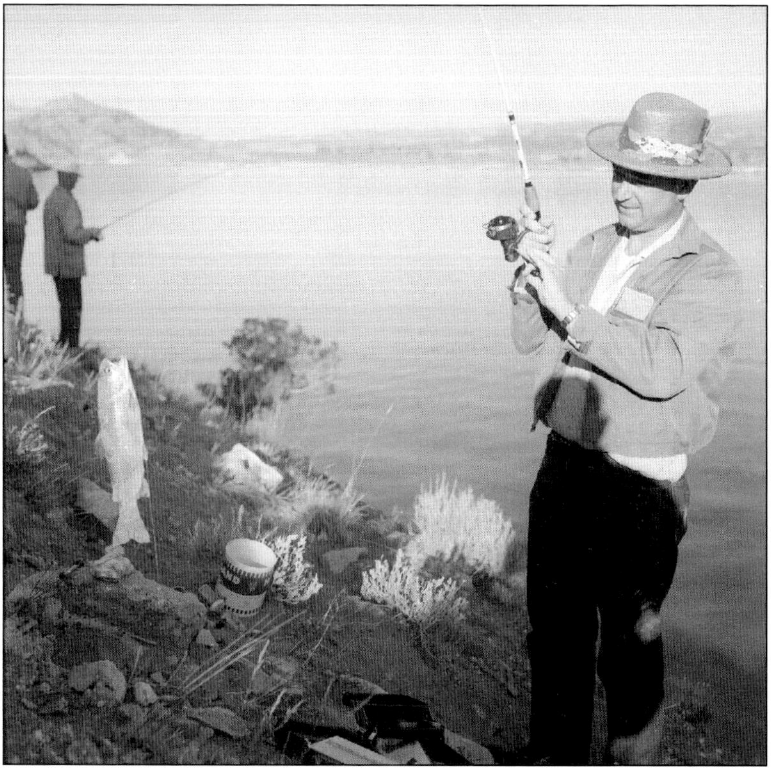

This happy fisherman and others in the background are fishing on the banks of the Flaming Gorge Reservoir. A successful day of fishing is evident, as this sportsman has caught a beautiful trout. (Courtesy of UCRHC.)

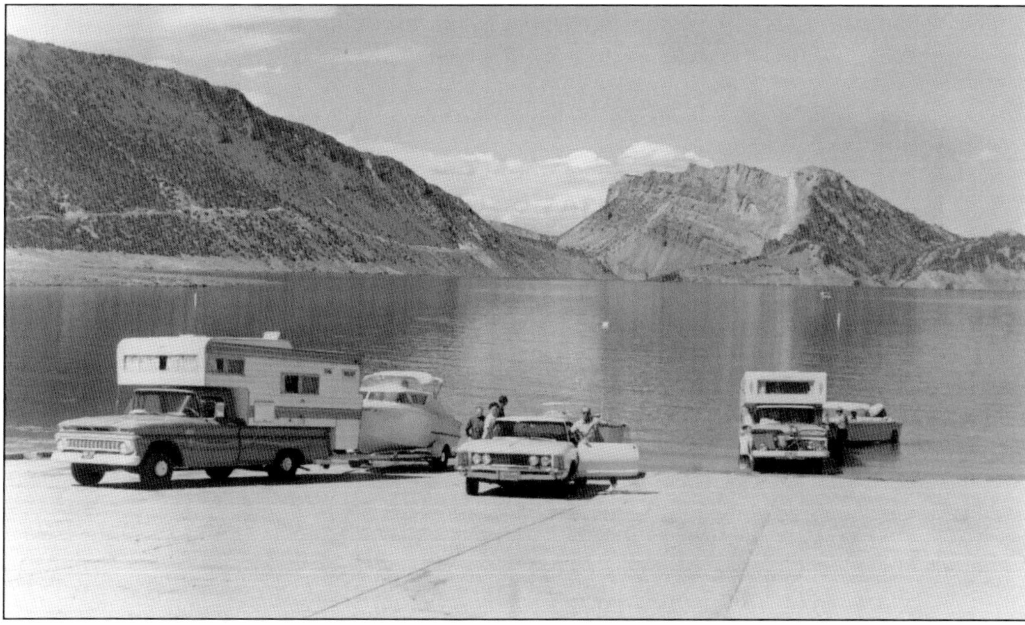

Visitors launch boats at the Antelope Boat Ramp at the Flaming Gorge National Recreation Area, Utah-Wyoming. Within the recreation area, nine concrete ramps have been constructed, enabling boaters to easily move their boats on and off the lake. (Courtesy of BOR.)

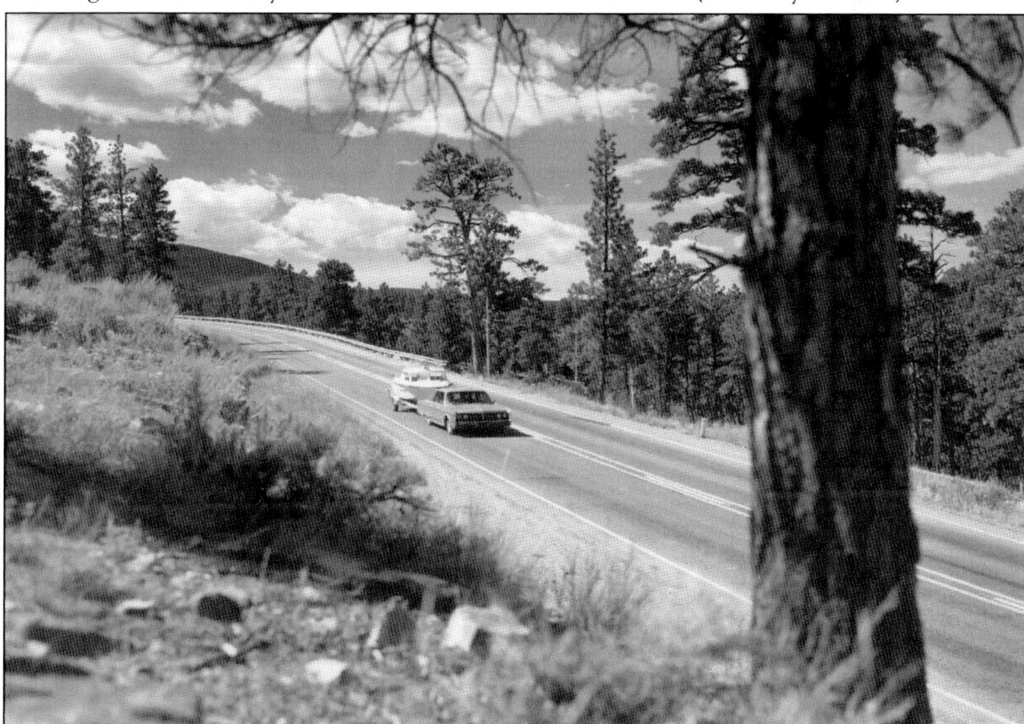

These recreationists are traveling along the scenic Manila Highway, eager to arrive at the nearest boat ramp, where they will launch their motorboat for a fun-filled day on the Flaming Gorge Reservoir. (Courtesy of USFS.)

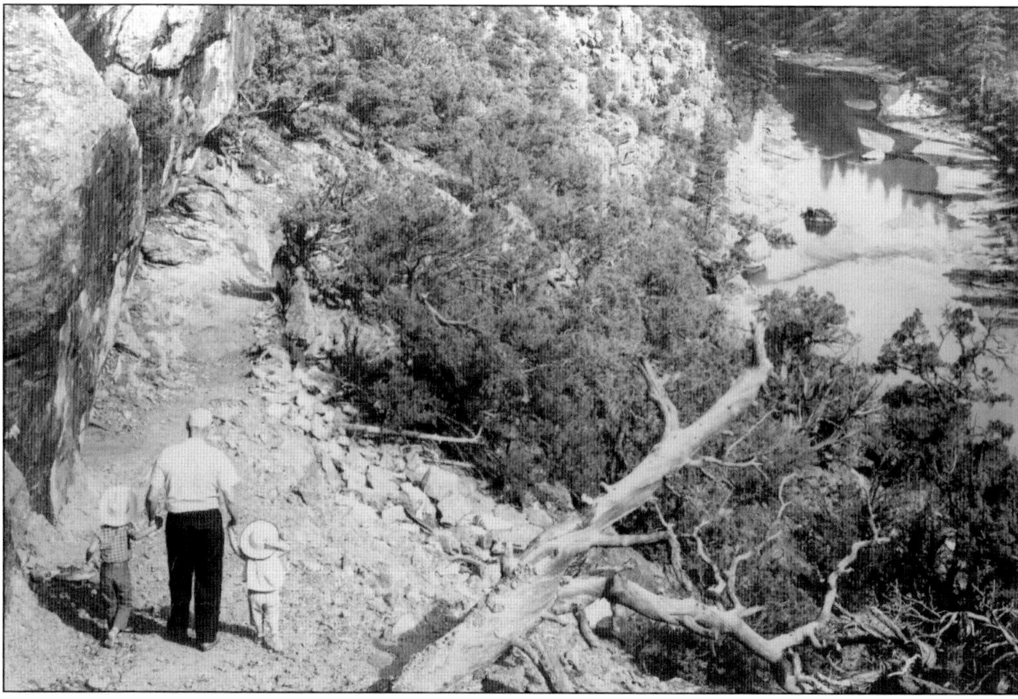

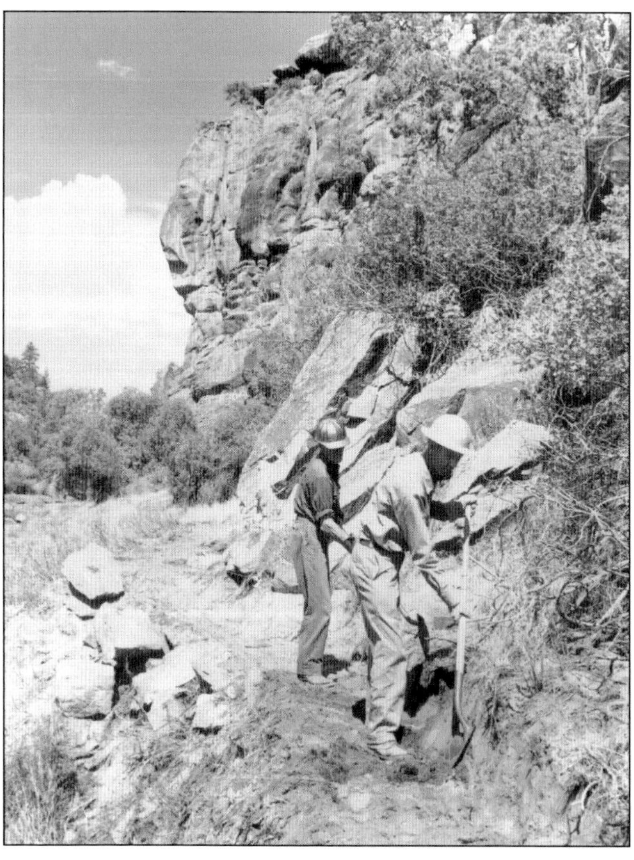

Gill McBride and his two children, Shari and Tommy, of Dutch John, Utah, explore the upper section of the trail being constructed by the US Forest Service. The trail will run parallel to the Green River, from a point about a half mile downstream from Flaming Gorge Dam to the Little Hole Area. (Courtesy of BOR.)

Victor West (left) and Ed Cross of North Carolina construct a section of the trail being built by the US Forest Service that will run parallel to the Green River. The trail, running through some very scenic canyon country, will offer a path for fishermen and nature lovers. The river, fed with water passing through the generating turbines at Flaming Gorge Dam, will maintain low temperatures, providing excellent trout fishing. (Courtesy of BOR.)

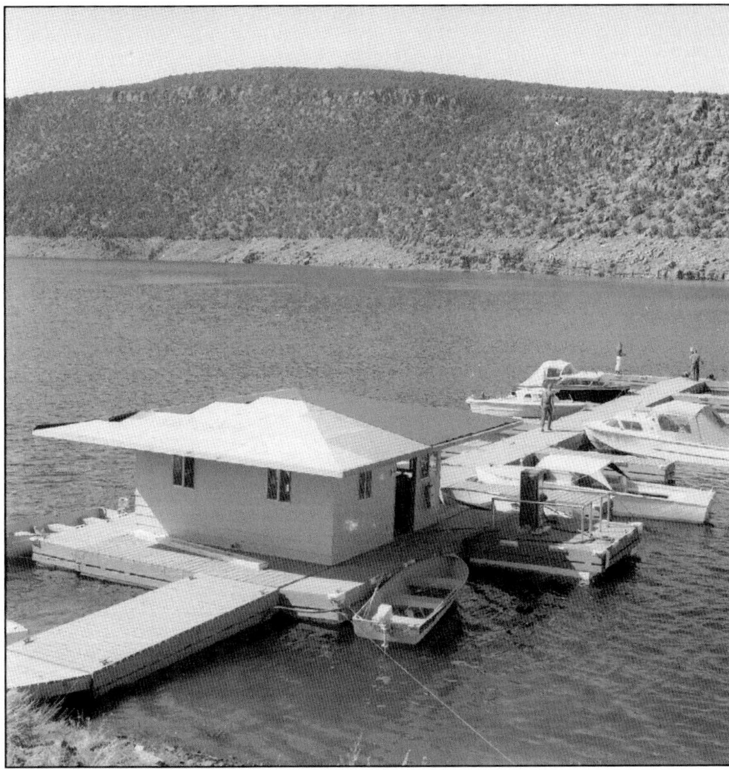

The new marina at Cedar Springs is seen here in use for the first time during the Labor Day holiday, September 7, 1964. The new facilities provided docking for visiting boatmen. Gas and oil, along with other fishing and boating supplies, were available. The enterprise is managed by a group of businessmen. (Courtesy of UCRHC.)

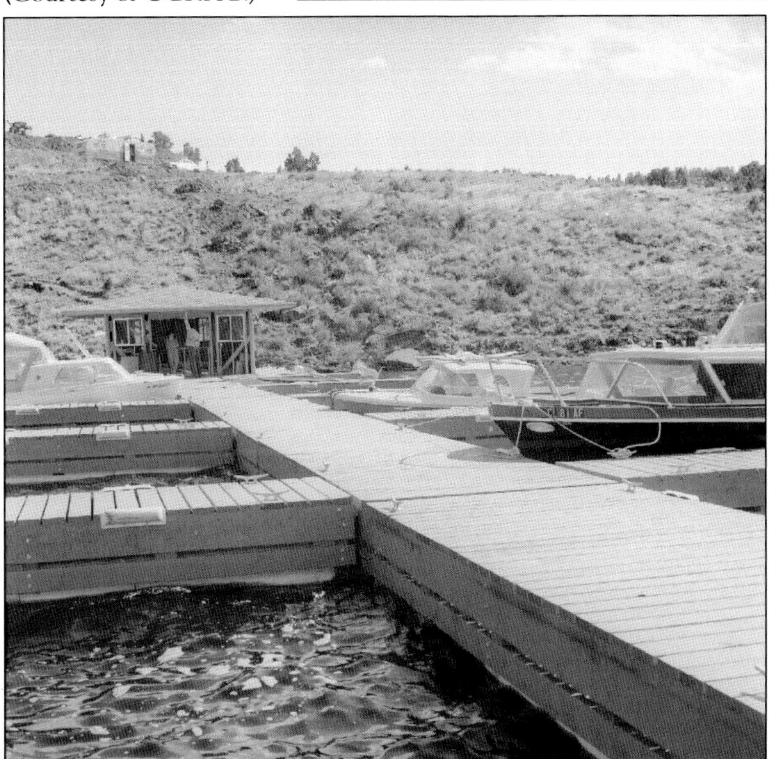

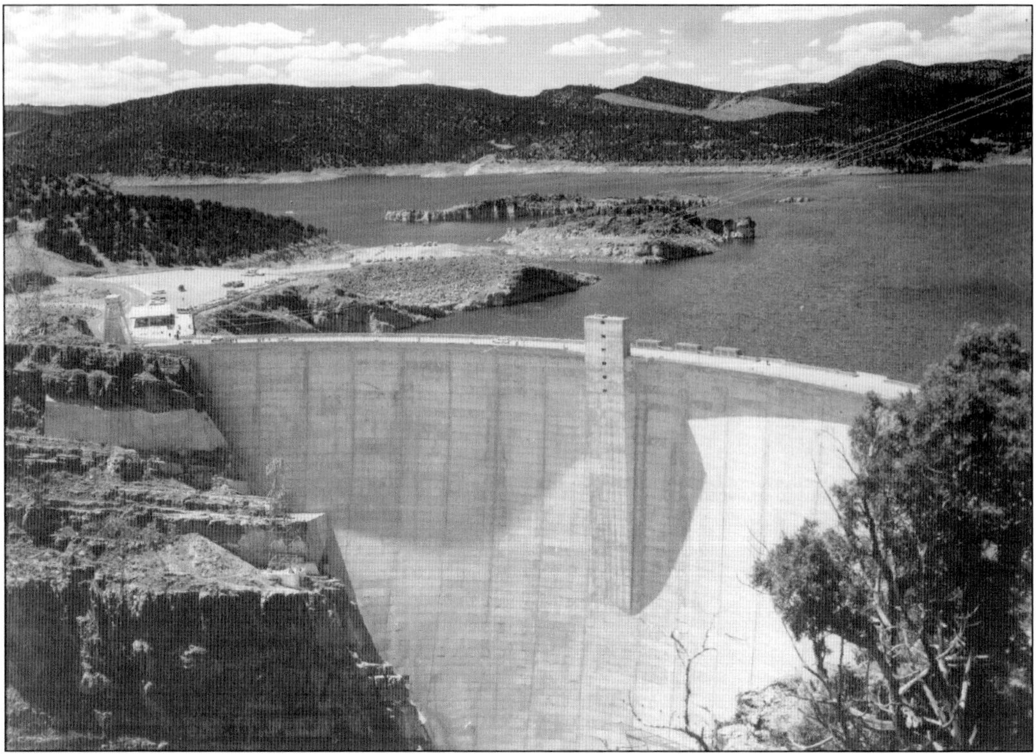

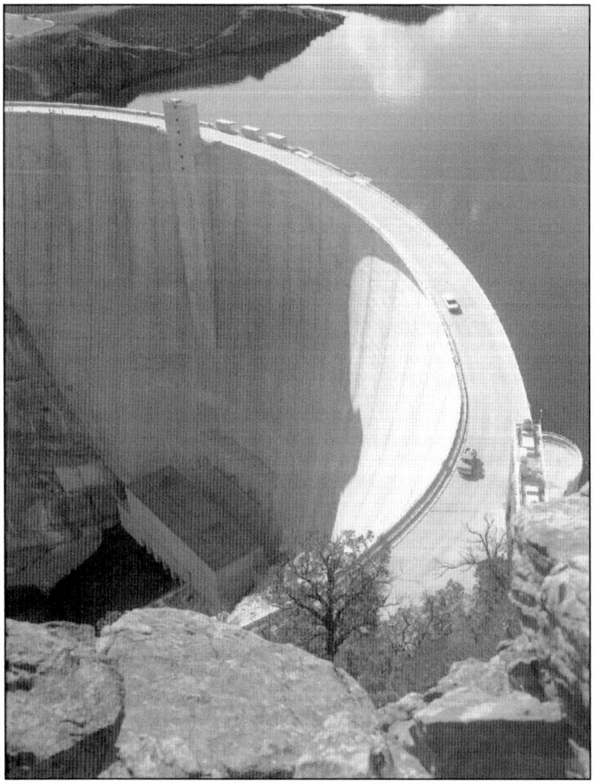

This is a scenic view of the Flaming Gorge Reservoir as it begins to fill after the completion of the Flaming Gorge Dam. The visitor center can be seen on the far left of the dam. (Courtesy of USFS.)

The highway across Flaming Gorge Dam is open to restricted travel. This view, looking toward the right abutment, shows the 1,285-foot crest on which the road is set. The roof of the powerhouse is visible in the lower portion of the photograph. The future promises many tourists traveling across this great wall of concrete. (Courtesy of UCRHC.)

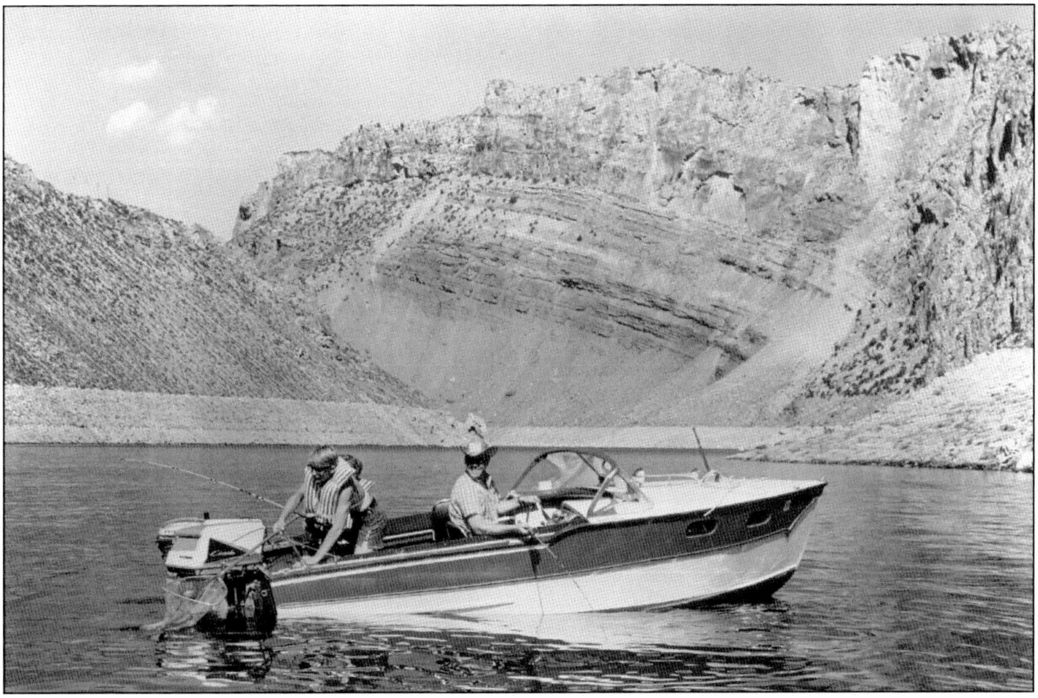

Flaming Gorge Reservoir provides many activities, including boating and fishing. A young boy and his father enjoy a day out on the waters of the gorge. (Courtesy of UCRHC.)

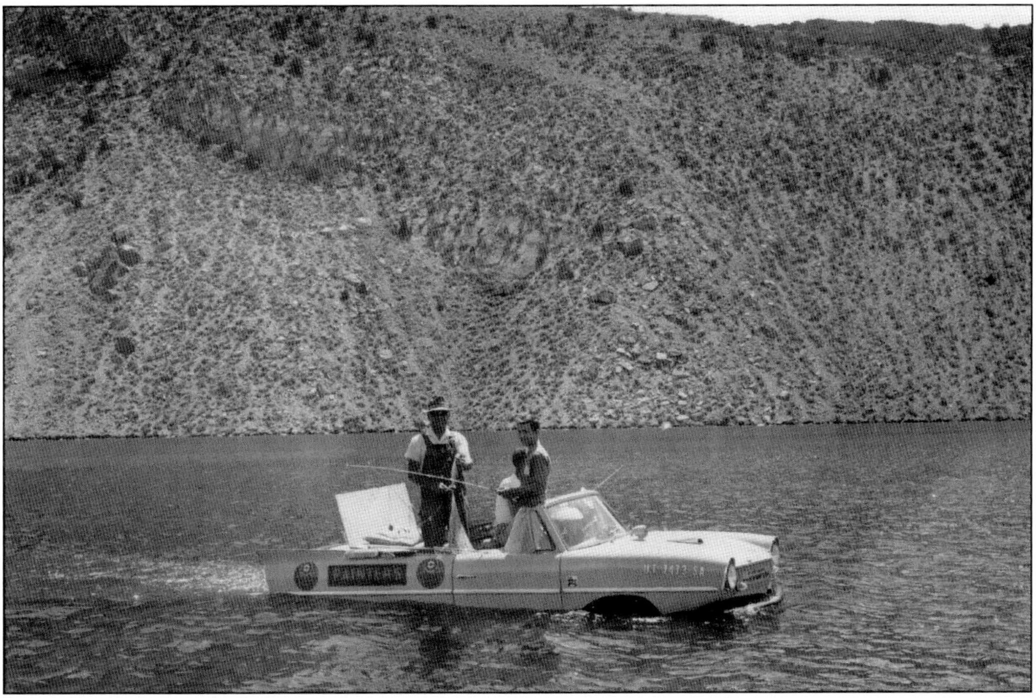

The Flaming Gorge Reservoir provides recreation for people from all over—in boats of all sorts of styles. Mr. Painter, the owner of A&W Root Beer in Salt Lake City, brought his unique car-boat to enjoy a day of boating and fishing. (Courtesy of USFS.)

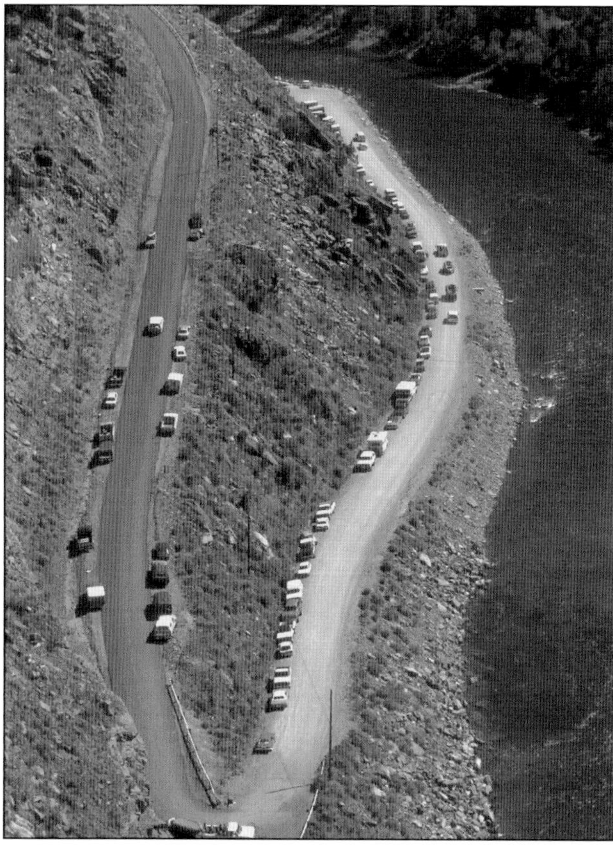

By July 1981, there was an abundant amount of "river runners" parked on the road leading to the Green River. They would leave their vehicles and be shuttled back after floating the Green River below the dam. (Courtesy of USFS.)

The white-water section of the Green River below the dam has been a popular place for avid anglers and adventurous rafters. Visitors can enjoy a day of scenic views, great fishing, and mild white-water rafting. (Courtesy of USFS.)

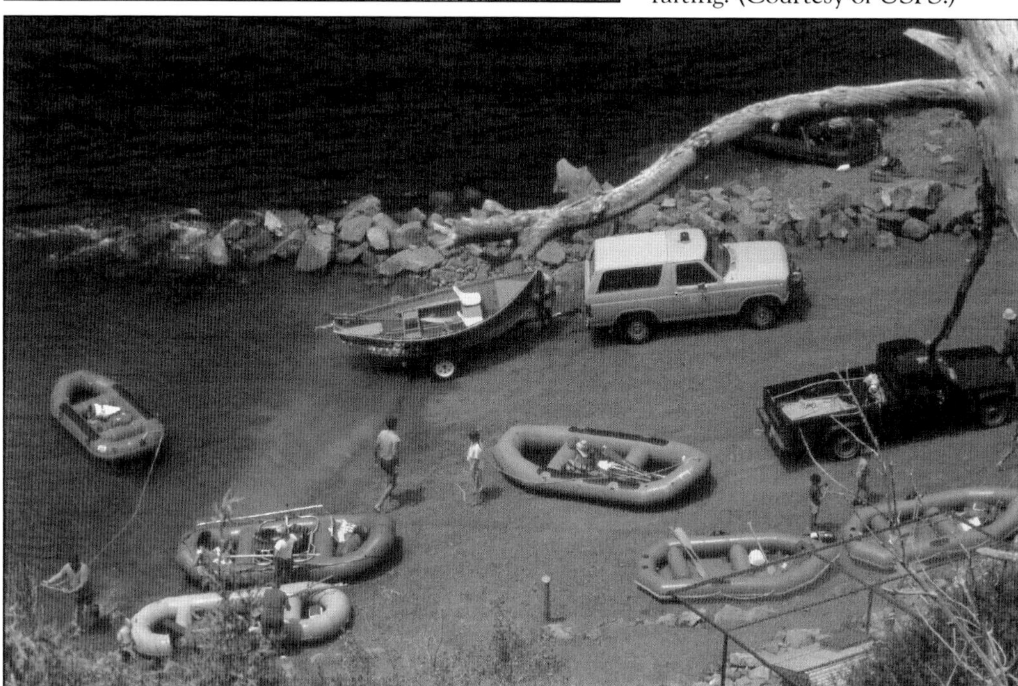

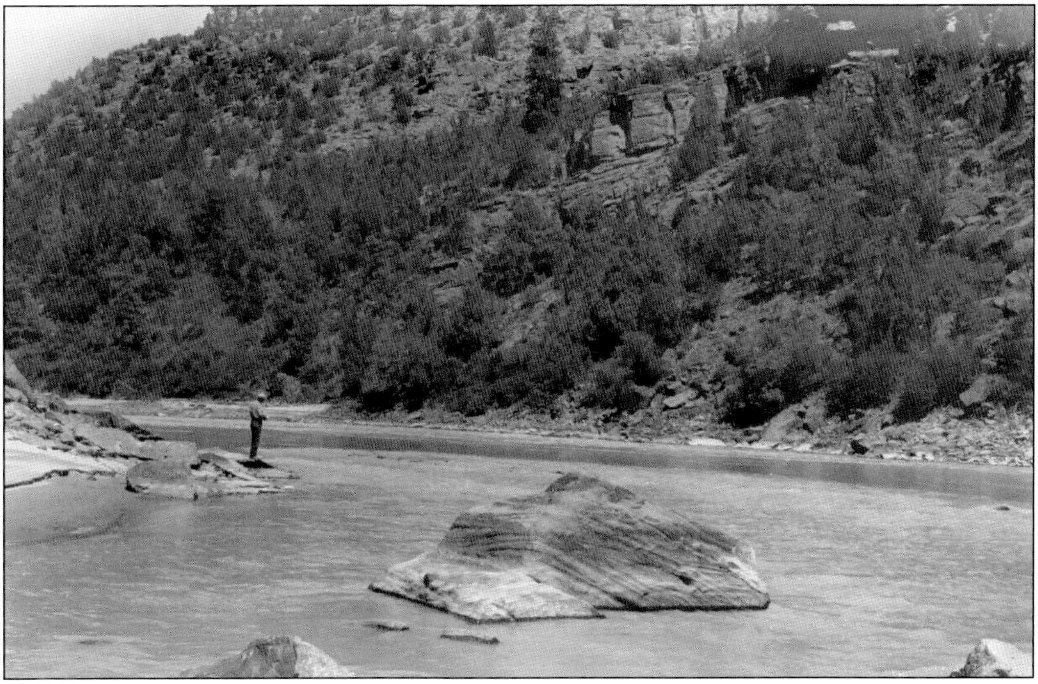

An early angler can be seen fly-fishing along the Green River below the Flaming Gorge Dam. Before the dam was built, the Green River was considered a general fishery, which meant that anglers could use any type of bait or means to catch fish. Since then, the Green River below the dam has been officially declared a "blue ribbon fishery," which limits the number and size of fish that can be kept. (Courtesy of UCRHC.)

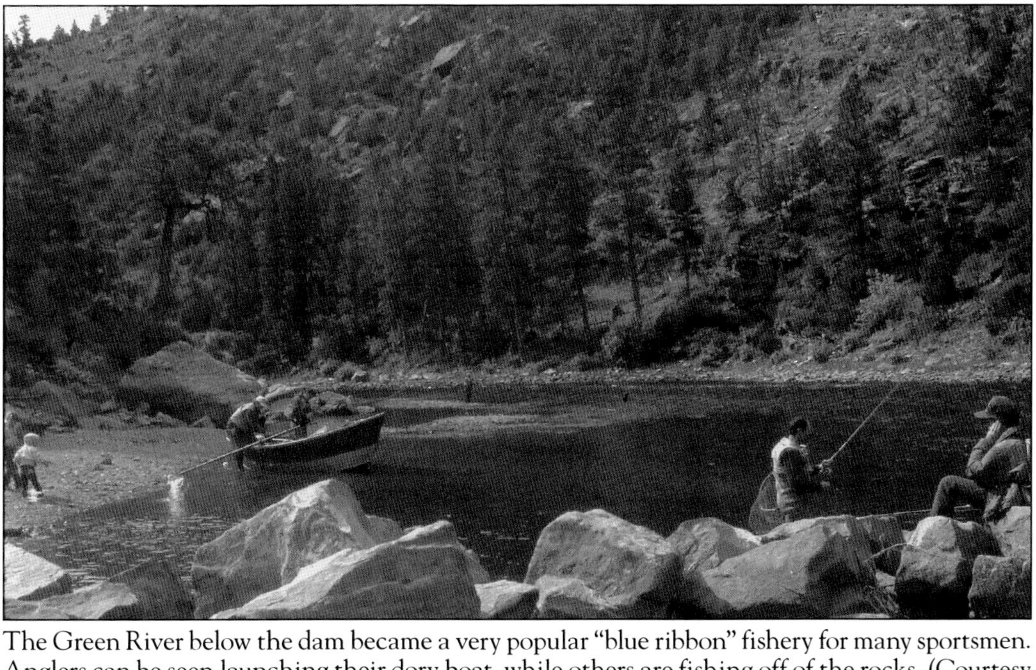

The Green River below the dam became a very popular "blue ribbon" fishery for many sportsmen. Anglers can be seen launching their dory boat, while others are fishing off of the rocks. (Courtesy of USFS.)

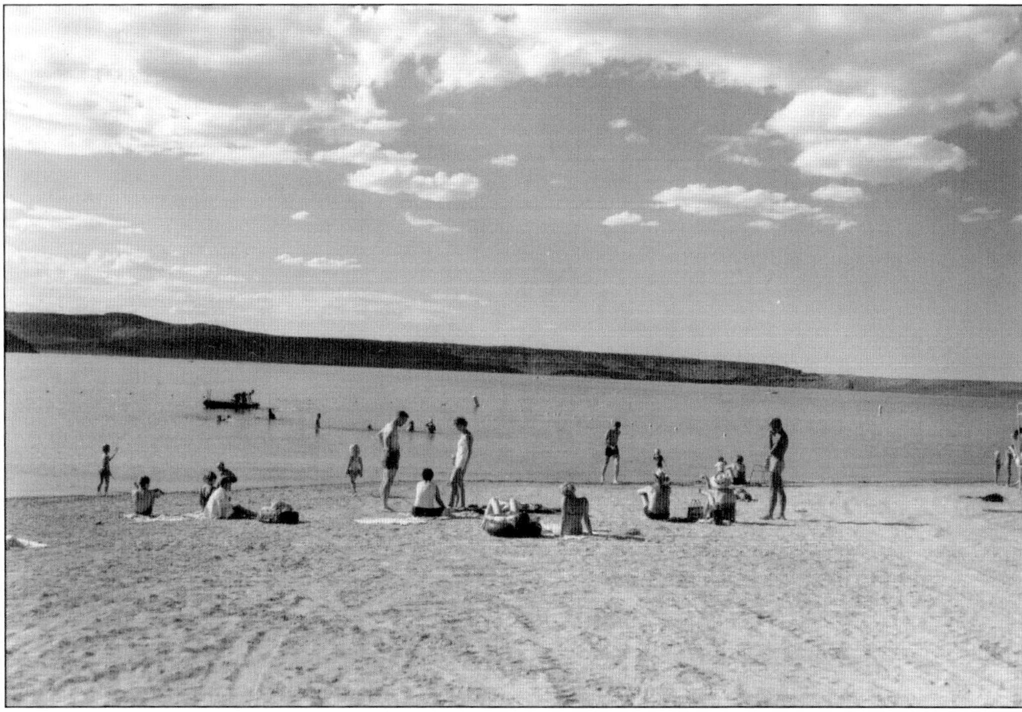

People enjoy a hot summer day on the shoreline. Swimmers and sunbathers bask in the sun at Flaming Gorge Reservoir, near the Manila, Utah, side. (Courtesy of USFS.)

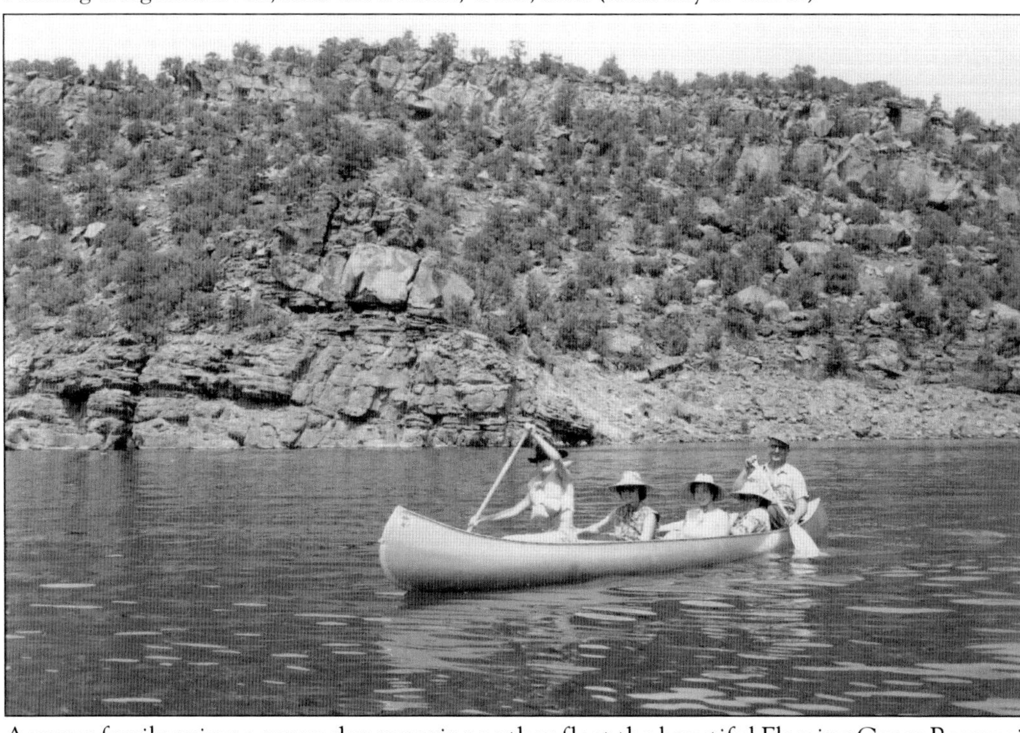

A young family enjoys a sunny day canoeing as they float the beautiful Flaming Gorge Reservoir opposite the mouth of Bear Creek. (Courtesy of USFS.)

Visitors stand on the boat dock at the Gooseneck Campground. This was one of many campgrounds built by the Ashley National Forest Service. These sites were conveniently built for boaters to access the campgrounds from the reservoir, allowing them to enjoy a simple picnic or adventure at an overnight camping area. (Courtesy of UCRHC.)

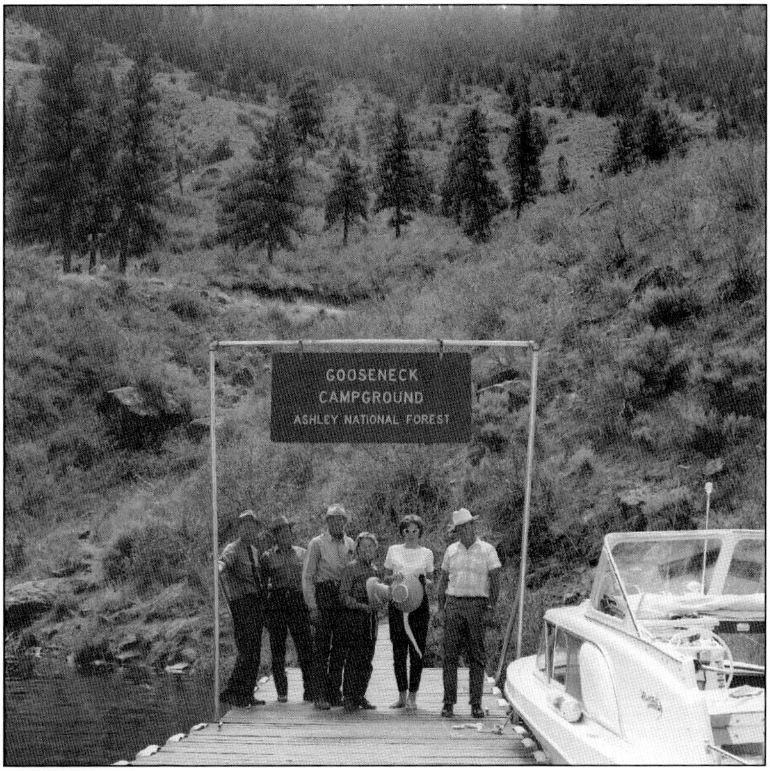

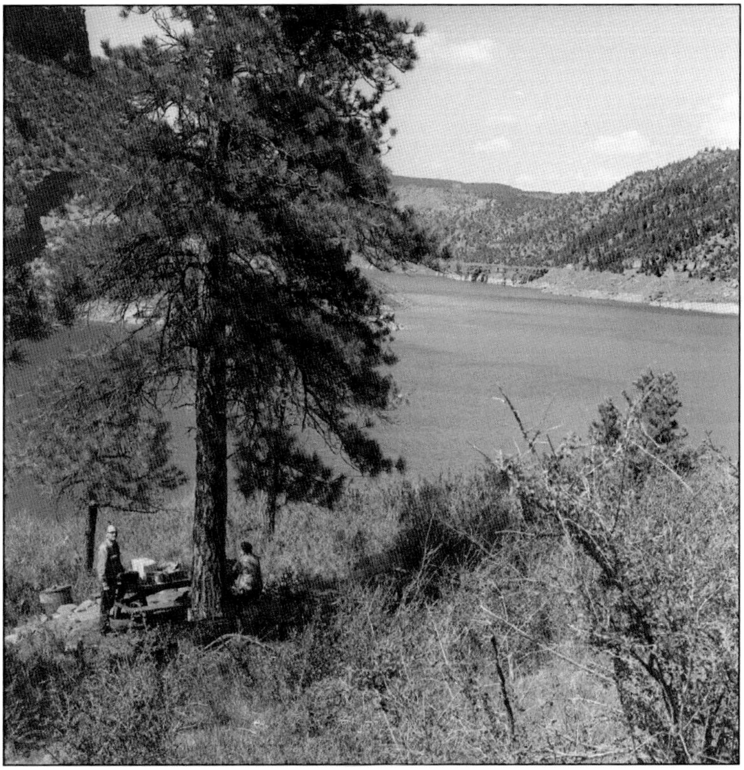

A couple enjoys an outdoor picnic and a panoramic view of the Flaming Gorge Reservoir near the Gooseneck Campground. (Courtesy of UCRHC.)

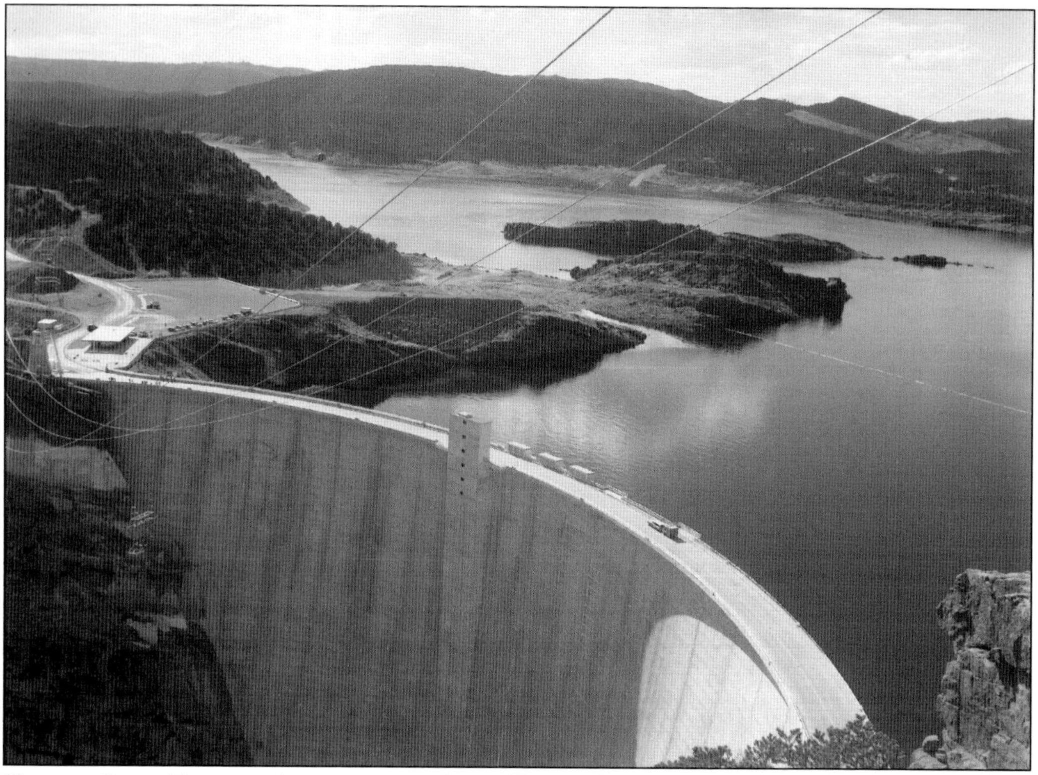

Flaming Gorge Reservoir brings visitors from all over the country to this beautiful scenic area. This aerial photograph shows a car crossing on the concrete road of Flaming Gorge Dam. The Flaming Gorge Visitor Center can be seen on the left. (Courtesy of UCRHC.)

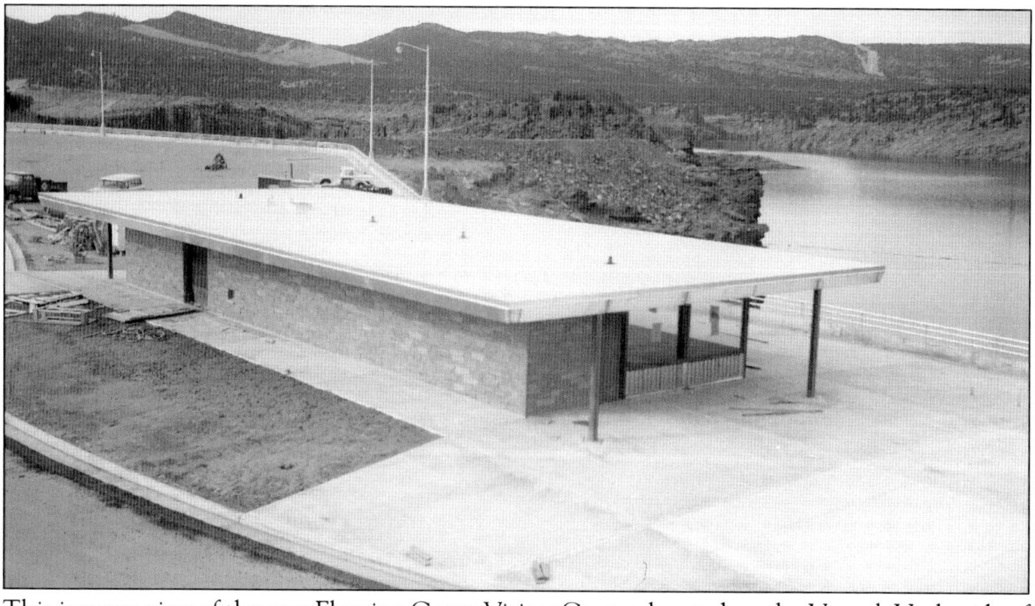

This is a rear view of the new Flaming Gorge Visitor Center, located on the Vernal, Utah, side of the approach road to Flaming Gorge Dam. Part of the reservoir can be seen in the background. The center's ample parking will be used during the tourism season. (Courtesy of UCRHC.)

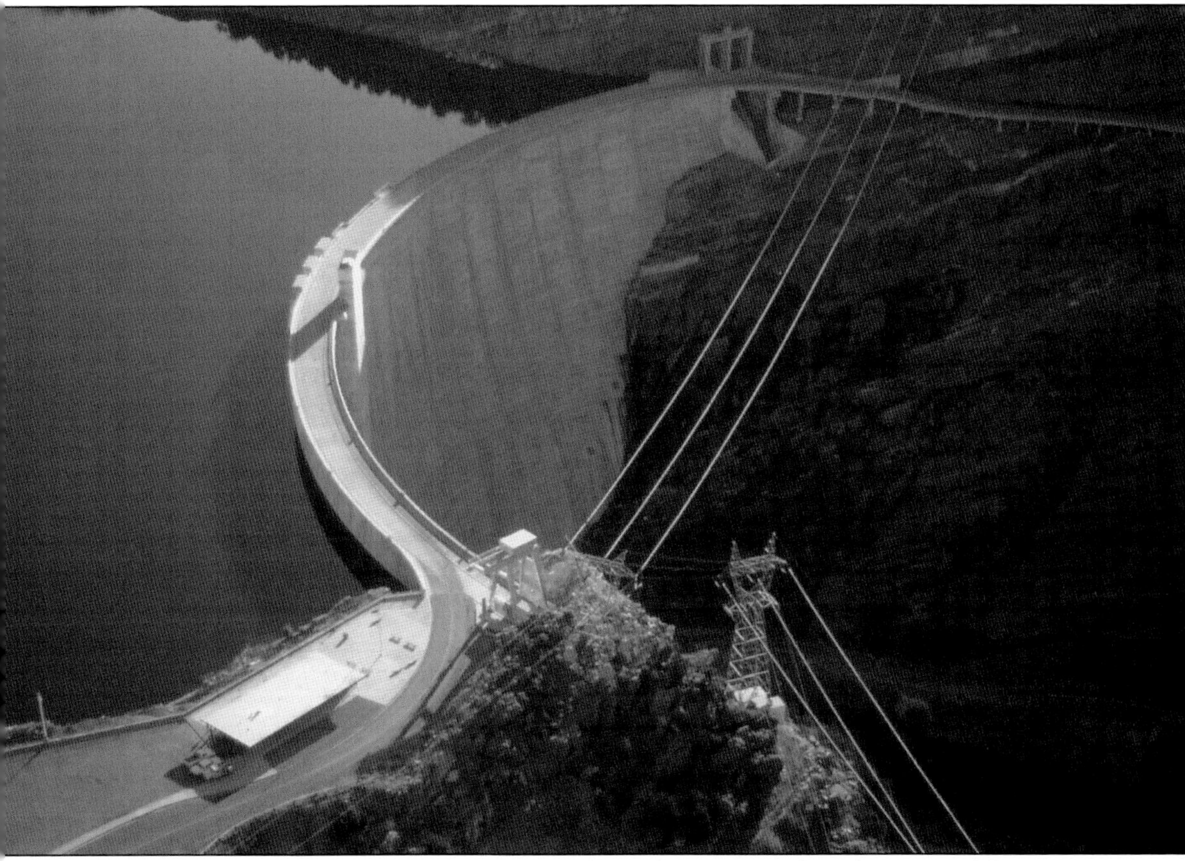

An aerial photograph of the Flaming Gorge Dam shows the precise arch architecture of this great concrete wall. Soon, visitors will traverse the curvy road to access other places in this beautiful recreational area. Many will stop at the newly built visitor center to seek education of the process and building of the majestic dam. Inside the center, many will even have the opportunity to take a tour of the dam, going inside the structure via an elevator. The building of the dam has provided unlimited recreation, such as boating, swimming, fishing, and camping. The US Forest Service administers the recreational facilities of the Flaming Gorge National Recreation Area and operates the Flaming Gorge Visitor Center in cooperation with the Bureau of Reclamation. (Courtesy of USFS.)

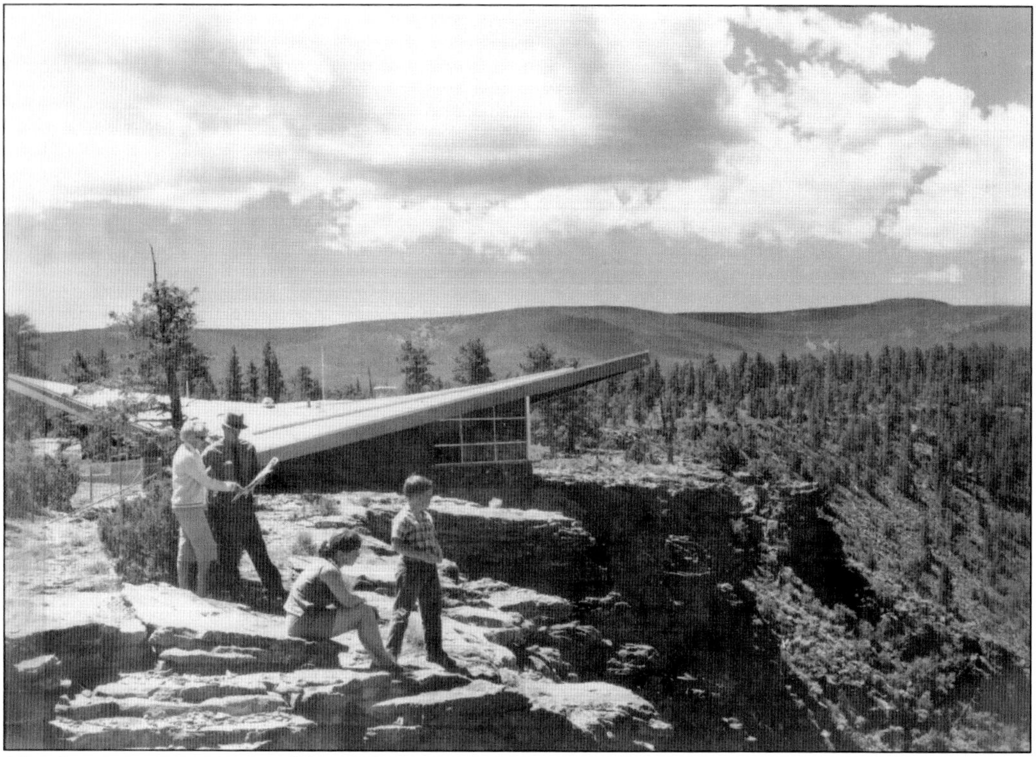

The dramatic wing shape of the Red Canyon Overlook and Visitor Center allows an unobstructed view of the deep canyon and the rugged cliffs overlooking the Green River. In the above photograph, visitors take in the beautiful scenic view of the canyon. This visitor center can be found near the Red Canyon Lodge. (Courtesy of USFS.)

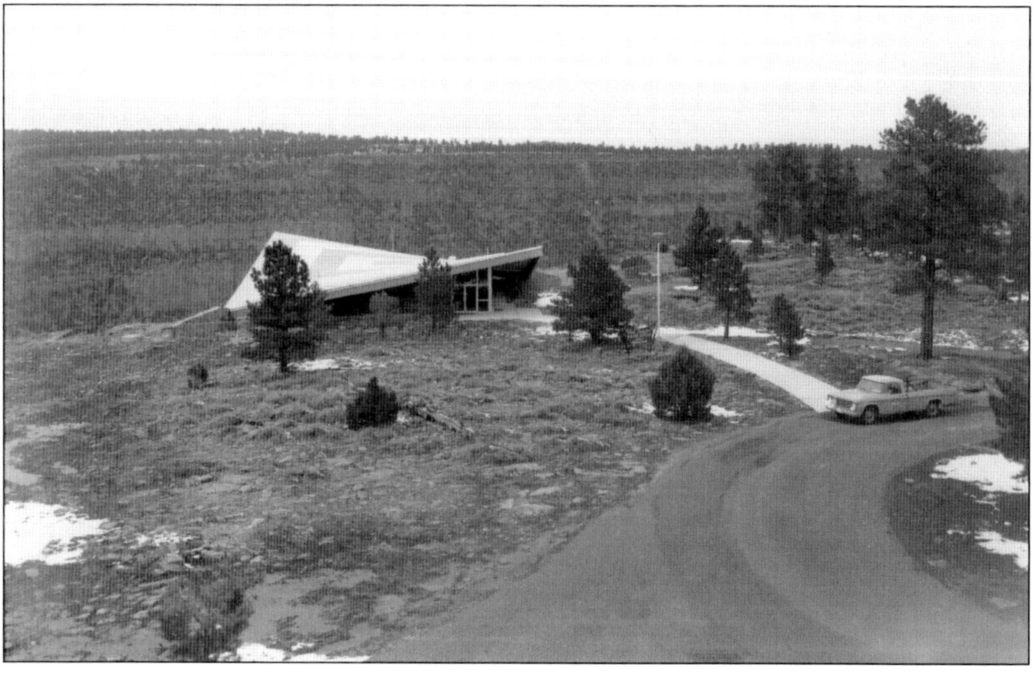

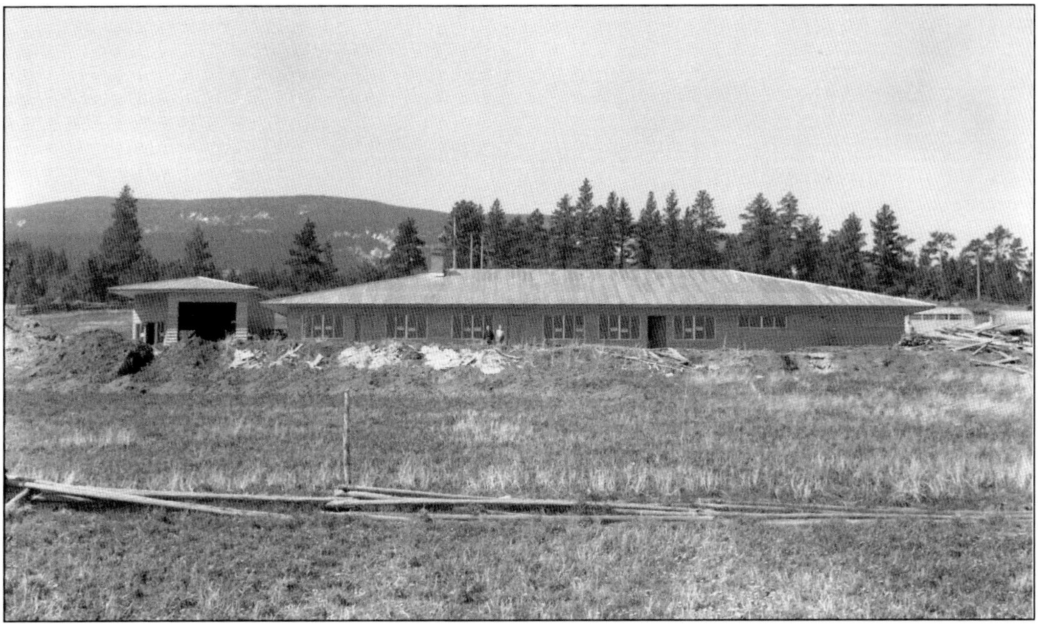

Flaming Gorge Lodge was built and owned by local contractor Francis Feltch. The grand opening took place at the end of May 1960. The lodge was leased and managed by Dwight L. Wetenkamp. This attractive roadhouse included a lounge, coffee shop, dining room, motel, swimming pool, store, service station, and dance floor. In these photographs, part of the facilities is still under construction. Completion of the project was expected by the end of that summer. The development is located 37 miles from Vernal and eight miles from Dutch John, Utah. Later, the business was purchased by Carl and Donna Collett. It is currently owned and operated by the Collett sons. (Courtesy of UCRHC.)

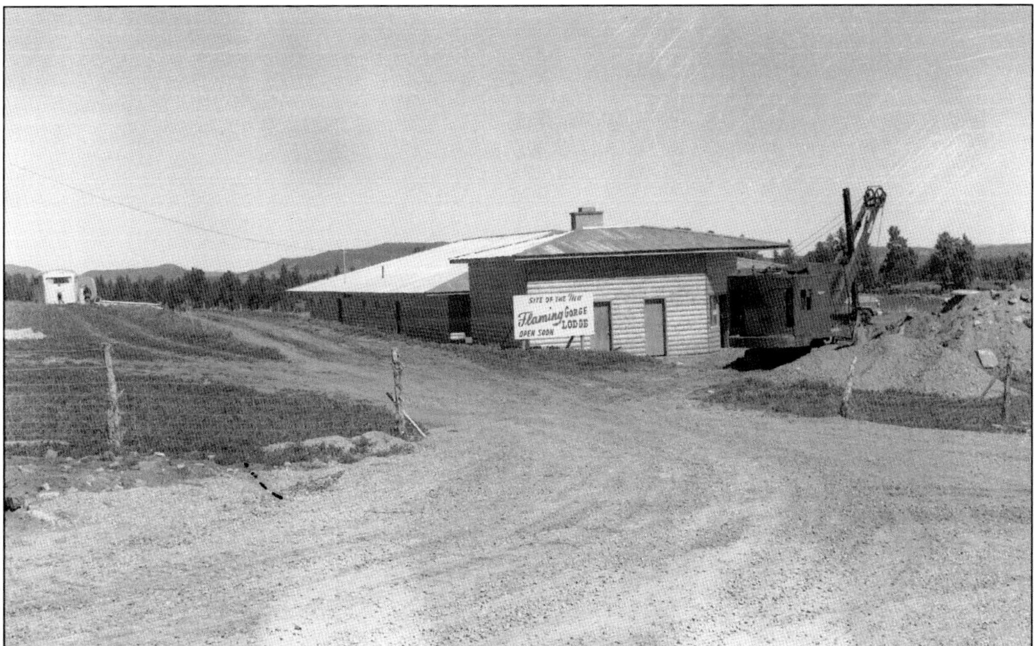